## IN THE SAME SERIES.

ON THE STUDY AND DIFFICULTIES OF MATHEMATICS. By Au-
GUSTUS DE MORGAN. Entirely new edition, with portrait of the author,
index, and annotations, bibliographies of modern works on algebra, the
philosophy of mathematics, pan-geometry, etc. Pp., 288. Cloth, $1.25 (5s.).

LECTURES ON ELEMENTARY MATHEMATICS. By JOSEPH LOUIS LA-
GRANGE. Translated from the French by *Thomas J. McCormack.* With
photogravure portrait of Lagrange, notes, biography, marginal analyses,
etc. Only separate edition in French or English. Pages, 172. Cloth,
$1.00 (5s.).

HISTORY OF ELEMENTARY MATHEMATICS. By DR. KARL FINK, late
Professor in Tübingen. Translated from the German by Prof. *Wooster
Woodruff Beman* and Prof. *David Eugene Smith.* (In preparation.)

THE OPEN COURT PUBLISHING CO.

324 DEARBORN ST., CHICAGO.

# MATHEMATICAL ESSAYS

AND

# RECREATIONS

BY

HERMANN SCHUBERT

PROFESSOR OF MATHEMATICS IN THE JOHANNEUM, HAMBURG, GERMANY

FROM THE GERMAN BY

THOMAS J. McCORMACK

---

CHICAGO

THE OPEN COURT PUBLISHING COMPANY

LONDON: KEGAN PAUL, TRENCH, TRUEBNER & CO.

1898

# TRANSLATOR'S NOTE.

THE mathematical essays and recreations in this volume are by one of the most successful teachers and text-book writers of Germany. The monistic construction of arithmetic, the systematic and organic development of all its consequences from a few thoroughly established principles, is quite foreign to the general run of American and English elementary text-books, and the first three essays of Professor Schubert will, therefore, from a logical and esthetic side, be full of suggestions for elementary mathematical teachers and students, as well as for non-mathematical readers. For the actual detailed development of the system of arithmetic here sketched, we may refer the reader to Professor Schubert's volume *Arithmetik und Algebra*, recently published in the Göschen-Sammlung (Göschen, Leipsic),— an extraordinarily cheap series containing many other unique and valuable text-books in mathematics and the sciences.

The remaining essays on "Magic Squares," "The Fourth Dimension," and "The History of the Squaring of the Circle," will be found to be the most complete generally accessible accounts in English, and to have, one and all, a distinct educational and ethical lesson.

In all these essays, which are of a simple and popular character, and designed for the general public, Professor Schubert has incorporated much of his original research.

THOMAS J. McCORMACK.

LA SALLE, Ill., December, 1898.

# CONTENTS.

# NOTION AND DEFINITION OF NUMBER.

MANY essays have been written on the definition of number. But most of them contain too many technical expressions, both philosophical and mathematical, to suit the non-mathematician. The clearest idea of what counting and numbers mean may be gained from the observation of children and of nations in the childhood of civilisation. When children count or add, they use either their fingers, or small sticks of wood, or pebbles, or similar things, which they adjoin singly to the things to be counted or otherwise ordinally associate with them. As we know from history, the Romans and Greeks employed their fingers when they counted or added. And even to-day we frequently meet with people to whom the use of the fingers is absolutely indispensable for computation.

Still better proof that the accurate association of such "other" things with the things to be counted is the essential element of numeration are the tales of travellers in Africa, telling us how African tribes sometimes inform friendly nations of the number of the enemies who have invaded their domain. The conveyance of the information is effected not by messengers, but simply by placing at spots selected for the purpose a number of stones exactly equal to the number of the invaders. No one will deny that the number of the tribe's foes is thus communicated, even though no name exists for this number in the languages of the tribes. The reason why the fingers are so universally employed as a means of numeration is, that every one possesses a definite number of fingers, sufficiently large for purposes of computation and that they are always at hand.

Besides this first and chief element of numeration which, as we have seen, is the exact, individual conjunction or association of other things with the things to be counted, is to be mentioned a second important element, which in some respects perhaps is not so absolutely essential; namely, that the things to be counted shall be regarded as of the same kind. Thus, any one who subjects apples and nuts collectively to a process of numeration will regard them for the time being as objects of the same kind, perhaps by subsuming them under the common notion of fruit. We may therefore lay down provisionally the following as a definition of counting : to count a group of things is to regard the things as the same in kind and to associate ordinally, accurately, and singly with them other things. In writing, we associate with the things to be counted simple signs, like points, strokes, or circles. The form of the symbols we use is indifferent. Neither need they be uniform. It is also indifferent what the spatial relations or dispositions of these symbols are. Although, of course, it is much more convenient and simpler to fashion symbols growing out of operations of counting on principles of uniformity and to place them spatially near each other. In this manner are produced what I have called * natural number-pictures ; for example,

Now-a-days such natural number-pictures are rarely employed, and are to be seen only on dominoes, dice, and sometimes, also, on playing-cards.

It can be shown by archæological evidence that originally numeral writing was made up wholly of natural number-pictures. For example, the Romans in early times represented all numbers, which were written at all, by assemblages of strokes. We have remnants of this writing in the first three numerals of the modern Roman system. If we needed additional evidence that the Romans originally employed natural number-signs, we might cite the passage in Livy, VII. 3, where we are told, that, in accordance with a very ancient law, a nail was annually driven into a certain spot in the sanctuary of

---

* *System der Arithmetik.* (Potsdam : Aug. Stein. 1885.)

Minerva, the "inventrix" of counting, for the purpose of showing the number of years which had elapsed since the building of the edifice. We learn from the same source that also in the temple at Volsinii nails were shown which the Etruscans had placed there as marks for the number of years.

Also recent researches in the civilisation of ancient Mexico show that natural number-pictures were the first stage of numeral notation. Whosoever has carefully studied in any large ethnographical collection the monuments of ancient Mexico, will surely have remarked that the nations which inhabited Mexico before its conquest by the Spaniards, possessed natural number-signs for all numbers from one to nineteen, which they formed by combinations of circles. If in our studies of the past of modern civilised peoples, we meet with natural number-pictures only among the Greeks or Romans, and some Oriental nations, the reason is that the other nations, as the Germans, before they came into contact with the Romans and adopted the more highly developed notation of the latter, were not yet sufficiently advanced in civilisation to feel any need of expressing numbers symbolically. But since the most perfect of all systems of numeration, the Hindu system of "local value," was introduced and adopted in Europe in the twelfth century, the Roman numeral system gradually disappeared, at least from practical computation, and at present we are only reminded by the Roman characters of inscriptions of the first and primitive stage of all numeral notation. To-day we see natural number-pictures, except in the above-mentioned games, only very rarely, as where the tally-men of wharves or warehouses make single strokes with a pencil or a piece of chalk, one for each bale or sack which is counted.

As in writing it is of consequence to associate with each of the things to be counted some simple sign, so in speaking it is of consequence to utter for each single thing counted some short sound. It is quite indifferent here what this sound is called, also whether the sounds which are associated with the things to be counted are the same in kind or not, and finally, whether they are uttered at equal or unequal intervals of time. Yet it is more convenient and simpler to employ the same sound and to observe equal intervals in their utterance. We arrive thus at natural number-words. For example, utterances like,

oh, oh-oh, oh-oh-oh, oh-oh-oh-oh, oh-oh-oh-oh-oh,

are natural number-words for the numbers from one to five. Number-words of this description are not now to be found in any known language. And yet we hear such natural number-words constantly, every day and night of our lives; the only difference being that the speakers are not human beings but machines—namely, the striking-apparatus of our clocks.

Word-forms of the kind described are too inconvenient, however, for use in language, not only for the speaker, on account of their ultimate length, but also for the hearer, who must be constantly on the *qui vive* lest he misunderstand a numeral word so formed. It has thus come about that the languages of men from time immemorial have possessed numeral words which exhibit no trace of the original idea of single association. But if we should always select for every new numeral word some new and special verbal root, we should find ourselves in possession of an inordinately large number of roots, and too severely tax our powers of memory. Accordingly, the languages of both civilised and uncivilised peoples always construct their words for larger numbers from words for smaller numbers. What number we shall begin with in the formation of compound numeral words is quite indifferent, so far as the idea of number itself is concerned. Yet we find, nevertheless, in nearly all languages one and the same number taken as the first station in the formation of compound numeral words, and this number is ten. Chinese and Latins, Fins and Malays, that is, peoples who have no linguistic relationship, all display in the formation of numeral words the similarity of beginning with the number ten the formation of compound numerals. No other reason can be found for this striking agreement than the fact that all the forefathers of these nations possessed ten fingers.

Granting it were impossible to prove in any other way that people originally used their fingers in reckoning, the conclusion could be inferred with sufficient certainty solely from this agreement with regard to the first resting-point in the formation of compound numerals among the most various races. In the Indo-Germanic tongues the numeral words from ten to ninety-nine are formed by composition from smaller numeral words. Two methods remain for continuing the formation of the numerals : either to take a new root as our basis of composition (hundred) or to go on counting from ninety-nine, saying tenty, eleventy, etc. If we were logically to follow out this second method we should get tenty-ty for a thousand, tenty-ty-ty for ten thousand, etc. But in the utterance of such words, the syllable *ty* would be so frequently repeated that the same inconvenience would be produced as above in our individual number-pictures. For this reason the genius which controls the formation of speech took the first course.

But this course is only logically carried out in the old Indian numeral words. In Sanskrit we not only have for ten, hundred, and thousand a new root, but new bases of composition also exist for ten thousand, one hundred thousand, ten millions, etc., which are in no wise related with the words for smaller numbers. Such roots exist among the Hindus for all numerals up to the number expressed by a one and fifty-four appended naughts. In no other language do we find this principle carried so far. In most languages the numeral words for the numbers consisting of a one with four and five appended naughts are compounded, and in further formations use is made of the words million, billion, trillion, etc., which really exhibit only one root, before which numeral words of the Latin tongue are placed.

Besides numeral word-systems based on the number *ten*, only logical systems are found based on the number five and on the number twenty. Systems of numeral words which have the basis five occur in equatorial Africa. (See the language-tables of Stanley's books on Africa.) The Aztecs and Mayas of ancient Mexico had the base twenty. In Europe it was mainly the Celts who reckoned with twenty as base. The French language still shows some few traces of the Celtic vicenary system, as in its word for eighty, *quatre-vingt*. The choice of five and of twenty as bases is explained simply enough by the fact that each hand has five fingers, and that hands and feet together have twenty fingers and toes.

As we see, the languages of humanity now no longer possess natural number-signs and number-words, but employ names and systems of notation adopted subsequently to this first stage. Accordingly, we must add to the definition of counting above given a third factor or element which, though not absolutely necessary, is yet important, namely, that we must be able to express the results of the above-defined associating of certain other things with the things to be counted, by some conventional sign or numeral word.

Having thus established what counting or *numbering* means, we are in a position to define also the notion of *number*, which we do by simply saying that by number we understand *the results* of counting or numeration. These are naturally composed of two elements. First, of the ordinary number-word or number-sign ; and secondly, of the word standing for the specific things counted. For example, eight men, seven trees, five cities. When, now, we have counted one group of things, and subsequently also counted another group of things of the same kind, and thereupon we conceive the two groups of things combined into a single group, we can save ourselves the labor of counting the things a third time by blending the number-pictures belonging to the two groups into a single number-picture belonging to the whole. In this way we arrive on the one hand at the idea of addition, and on the other, at the notion of "unnamed" number. Since we have no means of telling from the two original number-pictures and the third one which is produced from these, the kind or character of the things counted, we are ultimately led in our conception of number to abstract wholly from the nature of the things counted, and to form the definition of unnamed number.

We thus see that to ascend from the notion of named number to the notion of unnamed number, the notion of addition, joined to a high power of abstraction, is necessary. Here again our theory is best verified by observations of children learning to count and add. A child, in beginning arithmetic, can well understand what five pens or five chairs are, but he cannot be made to understand from this alone what five abstractly is. But if we put beside the first five pens three other pens, or beside the five chairs three other chairs, we can usually bring the child to see that five things plus three things are always eight things, no matter of what nature the things are, and that accordingly we need not always specify in counting what kind of things we mean. At first we always make the answer to our question of what five plus three is, easy for the child, by relieving him of the process of abstraction, which is necessary to ascend from the named to the unnamed number, an end which we accomplish by not asking first what five plus three is, but by associating with the numbers words designating things within the sphere of the child's experience, for example, by asking how many five pens plus three pens are.

The preceding reflexions have led us to the notion of unnamed or abstract numbers. The arithmetician calls these numbers positive whole numbers, or positive integers, as he knows of other kinds of numbers, for example, negative numbers, irrational numbers, etc. Still, observation of the world of actual facts, as revealed to us by our senses, can naturally lead us only to positive whole numbers, such only, and no others, being results of actual counting. All other kinds of numbers are nothing but artificial inventions of mathematicians created for the purpose of giving to the chief tool of the mathematician, namely, arithmetical notation, a more convenient and more practical form, so that the solution of the problems which arise in mathematics may be simplified. All numbers, excepting the results of counting above defined, are and remain mere symbols, which, although they are of incalculable value in mathematics, and, therefore, can scarcely be dispensed with, yet could, if it were a question of principle, be avoided. Kronecker has shown that any problem in which positive whole numbers are given, and only such are sought, always admits of solution without the help of other kinds of numbers, although the employment of the latter wonderfully simplifies the solution.

How these derived species of numbers, by the logical application of a single principle, flow naturally from the notion of number and of addition above deduced, I shall show in the next article entitled "Monism in Arithmetic."

# MONISM IN ARITHMETIC.

IN HIS *Primer of Philosophy*, Dr. Paul Carus defines monism as a "unitary conception of the world." Similarly, we shall understand by monism in a science the unitary conception of that science. The more a science advances the more does monism dominate it. An example of this is furnished by physics. Whereas formerly physics was made up of wholly isolated branches, like Mechanics, Heat, Optics, Electricity, and so forth, each of which received independent explanations, physics has now donned an almost absolute monistic form, by the reduction of all phenomena to the *motions* of molecules. For example, optical and electrical phenomena, we now know, are caused by the undulatory movements of the ether, and the length of the ether-waves constitutes the sole difference between light and electricity.

Still more distinctly than in physics is the monistic element displayed in pure arithmetic, by which we understand the theory of the combination of two numbers into a third by addition and the direct and indirect operations springing out of addition. Pure arithmetic is a science which has completely attained its goal, and which can prove that it has, exclusively by internal evidence. For it may be shown on the one hand that besides the seven familiar operations of addition, subtraction, multiplication, division, involution, evolution, and the finding of logarithms, no other operations are definable which present anything essentially new; and on the other hand that fresh extensions of the domain of numbers beyond irrational, imaginary, and complex numbers are arithmetically impossible. Arithmetic may be compared to a tree that has completed its growth, the boughs and branches of which may still increase in size or even give forth fresh sprouts, but whose main trunk has attained its fullest development.

Since arithmetic has arrived at its maturity, the more profound investigation of the nature of numbers and their combinations shows that a unitary conception of arithmetic is not only possible but also necessary. If we logically abide by this unitary conception, we arrive, starting from the notion of counting and the allied notion of addition, at all conceivable operations and at all possible extensions of the notion of number. Although previously expressed by Grassmann, Hankel, E. Schröder, and Kronecker, the author of the present article, in his "System of Arithmetic," Potsdam, 1885, was the first to work out the idea referred to, fully and logically and in a form comprehensible for beginners. This book, which Kronecker in his "Notion of Number," an essay published in Zeller's jubilee work, makes special mention of, is intended for persons proposing to learn arithmetic. As that cannot be the object of the readers of these essays, whose purpose will rather be the study of the logical construction of the science from some single fundamental principle, the following pages will simply give of the notions and laws of arithmetic what is absolutely necessary for an understanding of its development.

The starting-point of arithmetic is the idea of counting and of number as the result of counting. On this subject, the reader is requested to read the first essay of this collection. It is there shown that the idea of addition springs immediately from the idea of counting. As in counting it is indifferent in what order we count, so in addition it is indifferent, for the sum, or the result of the addition, whether we add the first number to the second or the second to the first. This law, which in the symbolic language of arithmetic, is expressed by the formula

$$a + b = b + a,$$

is called the *commutative law* of addition. Notwithstanding this law, however, it is evidently desirable to distinguish the two quantities which are to be summed, and out of which the sum is produced, by special names. As a fact, the two summands usually are distinguished in some way, for example, by saying $a$ is to be increased by $b$, or $b$ is to be added to $a$, and so forth. Here, it is plain, $a$ is always something that is to be increased, $b$ the increase. Accordingly it has been proposed to call the number which is regarded in addition as the passive number or the one to be changed, the *augend*, and the other which plays the active part, which accomplishes the change, so to speak, the *increment*. Both words are derived from the Latin and are appropriately chosen. Augend is derived from *augere*, to increase, and signifies that which is to be increased; increment comes from *increscere*, to grow, and signifies as in its ordinary meaning what is added.

Besides the commutative law one other follows from the idea of counting—the *associative law* of addition. This law, which has reference not to two but to three numbers, states that having a certain sum, $a + b$, it is indifferent for the result whether we increase the increment $b$ of that sum by a number, or whether we increase the sum itself by the same number. Expressed in the symbolic language of arithmetic this law reads,

$$a + (b + c) = (a + b) + c.$$

To obtain now all the rules of addition we have only to apply the two laws of commutation and association above stated, though frequently, in the deduction of the same rule, each must be applied many times. I may pass over here both the rules and their establishment.

In addition, two numbers, the augend $a$ and the increment $b$ are combined into a third number $c$, the sum. From this operation spring necessarily two inverse operations, the common feature of which is, that the sum sought in addition is regarded in both as known, and the difference that in the one the augend also is regarded as known, and in the other the increment. If we ask what number added to $a$ gives $c$, we seek the increment. If we ask what number increased by $b$ gives $c$, we seek the augend. As a matter of reckoning, the solution of the two questions is the same, since by the commutative law of addition $a + b = b + a$. Consequently, only one common name is in use for the two inverses of addition, namely, *subtraction*. But with respect to the notions involved, the two operations do differ, and it is accordingly desirable in a logical investigation of the structure of arithmetic, to distinguish the two by different names. As in all probability no terms have yet been suggested for these two kinds of subtraction, I propose here for the first time the following words for the two operations, namely, *detraction* to denote the finding of the increment, and *subtertraction* to denote the finding of the augend. We obtain these terms simply enough by thinking of the augmentation of some object already existing. For example, the cathedral at Cologne had in its tower an augend that waited centuries for its increment, which was only supplied a few decades ago. As the cathedral had originally a height of one hundred and thirty metres, but after completion was increased in height twenty-six metres, of the total height of one hundred and fifty-six metres one hundred and thirty metres is clearly the augend and twenty-six metres the increment. If, now, we wished to recover the augend we should have to pull down (Latin, *detrahere*) the upper part along the whole height. Accordingly, the finding of the augend is called *detraction*. If we sought the increment, we should have to pull out the original part from beneath (Latin, *subtertrahere*). For this reason, the finding of the increment is called *subtertraction*. Owing to the commutative law, the two inverse operations, as matters of computation, become one, which bears the name of *subtraction*. The sign of this operation is the minus sign, a horizontal stroke. The number which originally was sum, is called in subtraction minuend; the number which in addition was increment is now called detractor; the number which in addition was augend is now called subtertractor. Comprising the two conceptually different operations in one single operation, subtraction, we employ for the number which before was increment or augend, the term subtrahend, a word which on account of its passive ending is not very good, and for which, accordingly, E. Schröder proposes to substitute the word *subtrahent*, having an active ending. The result of subtraction, or what is the same thing, the number sought, is called the *difference*. The definition-formula of subtraction reads

$$a - b + b = a,$$

that is, $a$ minus $b$ is the number which increased by $b$ gives $a$, or the number which added to *b* gives *a*, according as the one or the other of the two operations inverse to addition is meant. From the formula for subtraction, and from the rules which hold for addition, follow now at once the rules which refer to both addition and subtraction. These rules we here omit.

From the foregoing it is plain that the minuend is necessarily larger than the subtrahent. For in the process of addition the minuend was the sum, and the sum grew out of the union of *two* natural number-pictures.* Thus 5 minus 9, or 11 minus 12, or 8 minus 8, are combinations of numbers *wholly destitute of meaning*; for no number, that is, no result of counting, exists that added to 9 gives the sum 5, or added to 12 gives the sum 11, or added to 8 gives 8. What, then, is to be done? Shall we banish entirely from arithmetic such meaningless combinations of numbers; or, since they have no meaning, shall we rather invest them with one? If we do the first, arithmetic will still be confined in the strait-jacket into which it was forced by the original definition of number as the result of counting. If we adopt the latter alternative we are forced to extend our notion of number. But in doing this, we sow the first seeds of the science of pure arithmetic, an organic body of knowledge which fructifies all other provinces of science.

What significance, then, shall we impart to the symbol

$$5 - 9?$$

Since 5 minus 9 possesses no significance whatever, we may, of course, impart to it any significance we wish. But as a matter of practical convenience it should be invested with no meaning that is likely to render it subject to exceptions. As the form of the symbol 5 — 9 is the form of a difference, it will be obviously convenient to give it a meaning which will allow us to reckon with it as we reckon with every other real difference, that is, with a difference in which the minuend is larger than the subtrahent. This being agreed upon, it follows at once that all such symbols in which the number before the minus sign is less than the number behind it by the same amount may be put equal to one another. It is practical,

---

* See page 2, *supra*.

therefore, to comprise all these symbols under some one single symbol, and to construct this latter symbol so that it will appear unequivocally from it by how much the number before the minus sign is less than the number behind it. This difference, accordingly, is written down and the minus sign placed before it.

If the two numbers of such a differential *form* are equal, a totally new sign must be invented for the expression of the fact, having no relation to the signs which state results of counting. This invention was not made by the ancient Greeks, as one might naturally suppose from the high mathematical attainments of that people, but by Hindu Brahman priests at the end of the fourth century after Christ. The symbol which they invented they called *tsiphra*, empty, whence is derived the English *cipher*. The form of this sign has been different in different times and with different peoples. But for the last two or three centuries, since the symbolic language of arithmetic has become thoroughly established as an international character, the form of the sign has been 0 (French *zéro*, German *null*).

In calling this symbol and the symbols formed of a minus sign followed by a result of counting, *numbers*, we widen the province of numbers, which before was wholly limited to results of counting. In no other way can zero and the negative numbers be introduced into arithmetic. No man can *prove* that 7 minus 11 is equal to 1 minus 5. Originally, both are meaningless symbols. And not until we agree to impart to them a significance which allows us to reckon with them as we reckon with real differences are we led to a statement of identity between 7 minus 11 and 1 minus 5. It was a long time before the negative numbers mentioned acquired the full rights of citizenship in arithmetic. Cardan called them, in his *Ars Magna*, 1545, *numeri ficti* (imaginary numbers), as distinguished from *numeri veri* (real numbers). Not until Descartes, in the first half of the seventeenth century, was any one bold enough to substitute *numeri ficti* and *numeri veri* indiscriminately for the same letter of algebraic expressions.

We have invested, thus, combinations of signs originally meaningless, in which a smaller number stood before than after a minus sign, with a meaning which enables us to reckon with such *apparent* differences exactly as we do with ordinary differences. Now it is just this practical shift of imparting meanings to combinations, which logically applied deduces naturally the whole system of arithmetic from the idea of counting and of addition, and which we may characterise, therefore, as the *foundation-principle* of its whole construction. This principle, which Hankel once called the *principle of permanence*, but which I prefer to call the PRINCIPLE OF NO EXCEPTION, may be stated in general terms as follows:

*In the construction of arithmetic every combination of two previously defined numbers by a sign for a previously defined operation ( plus, minus, times, etc.) shall be invested with meaning, even where the original definition of the operation used excludes such a combination; and the meaning imparted is to be such that the combination considered shall obey the same formula of definition as a combination having from the outset a signification, so that the old laws of reckoning shall still hold good and may still be applied to it.*

A person who is competent to apply this principle rigorously and logically will arrive at combinations of numbers whose results are termed irrational or imaginary with the same necessity and facility as at the combinations above discussed, whose results are termed negative numbers and zero. To think of such combinations as *results* and to call the products reached also "numbers" is a misuse of language. It were better if we used the phrase *forms of numbers* for all numbers that are not the results of counting. But *usus tyrannus!*

It will now be my task to show how all numbers at which arithmetic ever has arrived or ever can arrive naturally flow from the simple application of the principle of no exception.

Owing to the commutative and associative laws for addition it is wholly indifferent for the result of a series of additive processes in what order the numbers to be summed are added. For example,

$$a + (b + c + d) + (e + f) = (a + b + c) + (d + e) + f.$$

The necessary consequence of this is that we may neglect the consideration of the order of the numbers and give heed only to what the quantities are that are to be summed, and, when they are equal, take note of only two things, namely, of what the quantity which is to be repeatedly summed is called and how often it occurs. We thus reach the notion of multiplication. To multiply *a* by *b* means to form the sum of *b* numbers each of which is called *a*. The number conceived summed is called the multiplicand, the number which indicates or counts how often the first is conceived summed is called the multiplier.

It appears hence, that the multiplier must be a result of counting, or a number in the original sense of the word, but that the multiplicand may be any number hitherto defined, that is, may also be zero or negative. It also follows from this definition that though the multiplicand may be a concrete number the multiplier cannot. Therefore, the commutative law of multiplication does not hold when the multiplicand is concrete. For, to take an example, though there is sense in requiring four trees to be summed three times, there is no sense in conceiving the number three summed "four trees times." When, however, multiplicand and multiplier are unnamed results of counting, (abstract numbers,) two fundamental laws hold in multiplication, exactly analogous to the fundamental laws of addition, namely, the law of commutation and the law of association. Thus,

$$a \text{ times } b = b \text{ times } a,$$
$$\text{and, } a \text{ times } (b \text{ times } c) = (a \text{ times } b) \text{ times } c.$$

The truth and correctness of these laws will be evident, if keeping to the definition of multiplication as an abbreviated addition of equal summands, we go back to the laws of addition. Owing to the commutative law it is unnecessary, for purposes of practical reckoning, to distinguish multiplicand and multiplier. Both have, therefore, a common name: *factor*. The result of the multiplication is called the product; the symbol of multiplication is a dot (.) or a cross ($\times$), which is read "times." Joined with the fundamental formula above written are a group of subsidiary formulæ which give directions how a sum or difference is multiplied and how multiplication is performed with a sum or difference. I need not enter, however, into any discussion of these rules here.

As the combination of two numbers by a sign of multiplication has no significance according to our definition of multiplication, when the multiplier is zero or a negative number, it will be seen that we are again in a position where it is necessary to apply the above explained principle of no exception. We revert, therefore, to what we above established, that zero and negative numbers are symbols which have the form of differences, and lay down the rule that multiplications with zero and negative numbers shall be performed exactly as with real differences. Why, then, is minus one times minus one, for example, equal to plus one? For no other reason than that minus one can be multiplied with an ordinary difference, as, for example, 8 minus 5, by first multiplying by 8, then multiplying by 5, and subtracting the differences obtained, and because agreeably to the principle of no exception we must say that the multiplication must be performed according to exactly the same rule with a symbol which has the *form* of a difference whose minuend is less by one than its subtrahent.

As from addition two inverse operations, detraction and subtertraction, spring, so also from multiplication two inverse operations must proceed which differ from each other simply in the respect that in the one the multiplicand is sought and in the other the multiplier. As matters of computation, these two inverse operations coalesce in a single operation, namely, division, owing to the validity of the commutative law in multiplication. But in so far as they are different ideas, they must be distinguished. As most civilised languages distinguish the two inverse processes of multiplication in the case in which the multiplicand is a line, we will adopt for arithmetic a name which is used in this exception. Let us take this example,

$$4 \, yards \times 3 = 12 \, yards.$$

If twelve yards and four yards are given, and the multiplier 3 is sought, I ask, how many summands, each equal to four yards, give twelve yards, or, what is the same thing, how often I can lay a length of four yards on a length of twelve yards? But this is *measuring*. Secondly, if twelve yards and the number 3 are given, and the multiplicand four yards is sought, I ask what summand it is which taken three times gives twelve yards, or, what is the same thing, what part I shall obtain if I cut up twelve yards into three equal parts? But this is partition, or *parting*. If, therefore, the multi-

plier is sought we call the division *measuring*, and if the multiplicand is sought, we call it *parting*. In both cases the number which
was originally the product is called the dividend, and the result the
quotient. The number which originally was multiplicand is called
the measure; the number which originally was multiplier is called
the parter. The common name for measure and parter is divisor.
The common symbol for both kinds of division is a colon, a horizontal stroke, or a combination of both. Its definitional formula
reads,

$$(a \div b) \cdot b = a, \text{ or, } \frac{a}{b} \cdot b = a.$$

Accordingly, dividing $a$ by $b$ means, to find the number which multiplied by $b$ gives $a$, or to find the number *with* which $b$ must be
multiplied to produce $a$. From this formula, together with the
formulæ relative to multiplication, the well-known rules of division
are derived, which I here pass over.

In the dividend of a quotient only such numbers can have a
place which are the product of the divisor with some previously defined number. For example, if the divisor is 5 the dividend can
only be 5, 10, 15, and so forth, and 0, — 5, — 10 and so forth. Accordingly, a stroke of division having underneath it 5 and above it
a number different from the numbers just named is a combination
of symbols having no meaning. For example, $\frac{3}{5}$ or $\frac{12}{5}$ are meaningless symbols. Now, conformably to the principle of no exception
we must invest such symbols which have the form of a quotient
without their dividend being the product of the divisor with any
number yet defined, with a meaning such that we shall be able to
reckon with such apparent quotients as with ordinary quotients.
This is done by our agreeing always to put the product of such a
quotient form with its divisor equal to its dividend. In this way we
reach the definition of broken numbers or *fractions*, which by the
application of the principle of no exception spring from division exactly as zero and negative numbers sprang from subtraction. The
latter had their origin in the impossibility of the subtraction ; the
former have their origin in the impossibility of the division. Putting

together now both these extensions of the domain of numbers, we arrive at *negative fractional numbers.*

We pass over the easily deduced rules of computation for fractions and shall only direct the reader's attention to the connexion which exists between fractional and non-fractional or, as we usually say, whole numbers. Since the number 12 lies between the numbers 10 and 15, or, what is the same thing, $10 < 12 < 15$, and since $10 : 5 = 2$, $15 : 5 = 3$, we say also that $12 : 5$ lies between 2 and 3, or that

$$2 < \tfrac{12}{5} < 3.$$

In itself, the notion of "less than" has significance only for results of counting. Consequently, it must first be stated what is meant when it is said that 2 is less than $\tfrac{12}{5}$. Plainly, nothing is meant by this except that 2 times 5 is less than 12. We thus see that every broken number can be so interpolated between two whole numbers differing from each other only by 1 that the one shall be smaller and the other greater, where smaller and greater have the meaning above given.

From the above definitions and the laws of commutation and association all possible rules of computation follow, which in virtue of the principle of no exception now hold indiscriminately for all numbers hitherto defined. It is a consequence of these rules, again, that the combination of two such numbers by means of any of the operations defined must in every case lead to a number which has been already defined, that is, to a positive or negative whole or fractional number, or to zero. The sole exception is the case where such a number is to be divided by zero. If the dividend also is zero, that is, if we have the combination $\tfrac{0}{0}$, the expression is one which stands for any number whatsoever, because any number whatsoever, no matter what it is, if multiplied by zero gives zero. But if the dividend is not zero but some other number $a$, be it what it will, we get a quotient form to which *no* number hitherto defined can be equated. But we discover that if we apply the ordinary arithmetical rules to $a \div 0$ all such forms may be equated to one another both when $a$ is positive and also when $a$ is negative. We may therefore invent two new signs for such quotient forms, namely $+ \infty$ and

— ∞. We find, further, that in transferring the notions greater and less to these symbols, $+\infty$ is greater than any positive number, however great, and $-\infty$ is smaller than any negative number, however small. We read these new signs, accordingly, "plus infinitely great" and "minus infinitely great."

But even here arithmetic has not reached its completion, although the combination of as many previously defined numbers as we please by as many previously defined operations as we please will still lead necessarily to some previously defined number. Every science must make every possible advance, and still one step in advance is possible in arithmetic. For in virtue of the laws of commutation and association, which also fortunately obtain in multiplication, just as we advance from addition to multiplication, so here again we may ascend from multiplication to *an operation of the third degree*. For, just as for $a + a + a + a$ we read $4 . a$, so with the same reason we may introduce some more abbreviated designation for $a . a . a . a$. The introduction of this new operation is in itself simply a matter of convenience and not an extension of the ideas of arithmetic. But if after having introduced this operation we repeatedly apply the monistic principle of arithmetic, the principle of no exception, we reach new means of computation which have led to undreamt of advances not only in the hands of mathematicians but also in the hands of natural scientists. The abbreviated designation mentioned, which, fructified by the principle of no exception, can render science such incalculable services, is simply that of writing for a product of $b$ factors of which each is called $a$, $a^b$, which we read $a$ to the $b^{th}$ power. Here a new direct operation, that of *involution*, is defined, and from now on we are justified in distinguishing operations which are not inverses of others, as addition, multiplication, and involution, by *numbers of degree*. Addition is the direct operation of the first degree, multiplication that of the second degree, and involution that of the third degree. In the expression $a^b$ the passive number $a$ is called the *base*, the active number $b$ the *exponent*, the result, the *power*.

But whilst in the direct operations of the first and second degree, the laws of commutation and association hold, here in involution, the operation of third degree, the two laws are inapplicable, and the result of their inapplicability is that operations of a still higher degree than the third form no possible advancement of pure arithmetic. The product of $b$ factors $a$ is not equal to the product of $a$ factors $b$; that is, the law of commutation does not hold. The only two different integers for which $a$ to the $b^{th}$ power is equal to $b$ to the $a^{th}$ power are 2 and 4, for 2 to the 4$^{th}$ power is 16, and 4 to the second power also is 16. So, too, the law of association as a general rule does not hold. For it is hardly the same thing whether we take the $(b^c)^{th}$ power of $a$ or the $c^{th}$ power of $a^b$.

From the definition of involution follow the usual rules for reckoning with powers, of which we shall only mention one, namely, that the $(b-c)^{th}$ power of $a$ is equal to the result of the division of $a$ to the $b^{th}$ power by $a$ to the $c^{th}$ power. If we put here $c$ equal to $b$, we are obliged, by the principle of no exception, to put $a$ to the $0^{th}$ power equal to 1; a new result not contained in the original notion of involution, for that implied necessarily that the exponent should be a result of counting. Again, if we make $b$ smaller than $c$ we obtain a *negative exponent*, which we should not know how to dispose of if we did not follow our monistic law of arithmetic. According to the latter, $a$ to the $(b-c)^{th}$ power must still remain equal to $a^b$ divided by $a^c$ even when $b$ is smaller than $c$. Whence follows that $a$ to the minus $d^{th}$ power is equal to 1 divided by $a$ to the $d^{th}$ power, or to take specific numbers, that 3 to the minus 2$^{nd}$ power is equal to $\frac{1}{9}$.

At this point, perhaps, the reader will inquire what $a$ raised to a fractional power is. But this can be explained only when we have discussed the inverse processes of involution, to which we now pass.

If $a^b = c$, we may ask two questions: first, what the base is which raised to the $b^{th}$ power gives $c$; the second, what the exponent of the power is to which $a$ must be raised to produce $c$. In the first case we seek the base, and term the operation which yields this result *evolution*; in the second case we seek the exponent and call the operation which yields this exponent, the *finding of the logarithm*. In the first case, we write $\sqrt[b]{c} = a$ (which we read, the $b^{th}$ root of $c$ is equal to $a$), and call $c$ the *radicand*, $b$ the *exponent of the root*, and $a$

the *root*. In the second case, we write $\log_a c = b$ (which we read, the logarithm of $c$ to the base $a$ is equal to $b$), and call $c$ the *logarithmand* or *number*, $a$ the *base of the logarithm*, and $b$ the *logarithm*.

While, owing to the validity of the law of commutation in addition and multiplication, the two inverse processes of those operations are identical so far as computation is concerned, here in the case of involution the two inverse operations are in this regard essentially different, for in this case the law of commutation does not hold.

From the definitional formulæ for evolution and the finding of logarithms, namely,

$$(\sqrt[b]{c})^b = c, \text{ and } (a)^{\log_a c} = c,$$

follow, by the application of the laws of involution, the rules for computation with roots and logarithms. These rules we pass over here, only remarking, first, that for the present $\sqrt[b]{c}$ has meaning only when $c$ is the $b^{th}$ power of some number already defined ; and, secondly, that for the present also $\log_a c$ has meaning only when $c$ can be produced by raising the number $a$ to some power which is a number already defined. In the phrase ''has only meaning for the present'' is contained a possibility of new extensions of the domain of number. But before we pass to those extensions we shall first make use of the idea of evolution just defined to extend the notion of power also to cases in which the exponent is a fractional number.

According to the original definition of involution, $a^b$ was meaningless except where $b$ was a result of counting. But afterwards, even powers which had for their exponents zero or a negative integer could be invested with meaning. Now we have to consider the arithmetical combination '' $a$ raised to the fractional power $\frac{p}{q}$.'' The principle of no exception compels us to give to the arithmetical combination '' $a$ to the $\frac{p}{q}^{th}$ power '' a significance such that all the rules of computation will hold with respect to it. Now, one rule that holds is, that the $m^{th}$ power of the $n^{th}$ power of $a$ is equal to the $(m \times n)^{th}$ power of $a$. Consequently, the $q^{th}$ power of $a$ raised to the $\frac{p}{q}^{th}$ power must be equal to $a$ raised to a power whose exponent is equal to $\frac{p}{q}$ times $q$. But the last-mentioned product gives, according to the definition of division, the number $p$. Consequently the symbol $a$ to the $\frac{p}{q}$ th power is so constituted that its $q^{th}$ power is equal to $a$ to the $p^{th}$ power; i. e., it is equal to the $q^{th}$ root of $a^p$. Similarly, we find that the symbol "$a$ to the minus $\frac{p}{q}$ th power" must be put equal to 1 divided by the $q^{th}$ root of $a$ to the $p^{th}$ power, if we are to reckon with this symbol as we do with real powers. Again, just as $a$ to the $b^{th}$ power is invested with meaning when $b$ is a fractional number, so some meaning harmonious with the principle of no exception must be imparted to the $b^{th}$ root of $c$ where $b$ is a positive or negative fractional number. For example, the three-fourths$^{th}$ root of 8 is equal to 8 to the $\frac{4}{3}$ power, that is, to the cube root of 8 to the $4^{th}$ power, or 16.

The principle underlying arithmetic now also compels us to give to the symbol the "$b^{th}$ root of $c$" a meaning when $c$ is not the $b^{th}$ power of any number yet defined. First, let $c$ be any *positive* integer or fraction. Then always to be able to reckon with the $b^{th}$ root of $c$ in the same way that we do with extractible roots, we must agree always to put the $b^{th}$ power of the $b^{th}$ root of $c$ equal to $c$—for example, $(\sqrt[2]{3})^2$ always exactly equal to 3. A careful inspection of the new symbols, which we will also call numbers, shows, that though no one of them is exactly equal to a number hitherto defined, yet by a certain extension of the notions greater and less, two numbers of the character of numbers already defined may be found for each such new number, such that the new number is greater than the one and less than the other of the two, and that further, these two numbers may be made to differ from each other by as small a quantity as we please. For example,

$$(\tfrac{7}{5})^3 = \tfrac{343}{125} = 2\tfrac{93}{125} < 3 < 3\tfrac{3}{8} = \tfrac{27}{8} = (\tfrac{3}{2})^3.$$

The number 3, as we see, is here included between two limits which are the third powers of two numbers $\frac{7}{5}$ and $\frac{3}{2}$ whose difference is $\frac{1}{10}$. We could also have arranged it so that the difference should be equal to $\frac{1}{100}$, or to any specified number, however small. Now, instead of putting the symbol "less than" between $(\tfrac{7}{5})^3$ and 3, and between 3 and $(\tfrac{3}{2})^3$, let us put it between their third roots; for example, let us say:

$$\tfrac{7}{5} < \sqrt[3]{3} < \tfrac{3}{2}, \text{ meaning by this that } (\tfrac{7}{5})^3 < 3 < (\tfrac{3}{2})^3.$$

In this sense we may say that the new numbers always lie *between*

two old numbers whose difference may be made as small as we
please. Numbers possessing this property are called *irrational* numbers, in contradistinction to the numbers hitherto defined, which are
termed *rational*. The considerations which before led us to negative
rational numbers, now also lead us to negative irrational numbers.
The repeated application of addition and multiplication as of their
inverse processes to irrational numbers, (numbers which though not
exactly equal to previously defined rational numbers may yet be
brought as near to them as we please,) again simply leads to numbers of the same class.

A totally new domain of numbers is reached, however, when we
attempt to impart meaning to *the square roots of negative numbers*.
The square root of minus 9 is neither equal to plus 3 nor to minus
3, since each multiplied by itself gives plus 9, nor is it equal to any
other number hitherto defined. Accordingly, the square root of minus
9 is a new number-form, to which, harmoniously with the principle
of no exception, we may give the definition that $(\sqrt[2]{-9})^2$ shall always be put equal to minus 9.* Keeping to this definition we see
at once that $\sqrt{-a}$, where $a$ is any positive rational or irrational
number, is a symbol which can be put equal to the product of $\sqrt{+a}$
by $\sqrt{-1}$. In extending to these new numbers the rights of arithmetical citizenship, in calling them also "numbers," and so shaping
their definition that we can reckon with them by the same rules as
with already defined numbers, we obtain a fourth extension of the
domain of numbers which has become of the greatest importance
for the progress of all branches of mathematics. The newly defined
numbers are called *imaginary*, in contradistinction to all heretofore
defined, which are called *real*. Since all imaginary numbers can be
represented as products of real numbers with the square root of
minus one, it is convenient to introduce for this one imaginary number some concise symbol. This symbol is the first letter of the word
imaginary, namely, $i$; so that we can always put for such an expression as $\sqrt{-9}$, $3 \cdot i$.

If we combine real and imaginary numbers by operations of the

---

* Henceforward we shall use the simpler sign $\sqrt{\phantom{-}}$ for $\sqrt[2]{\phantom{-}}$.

first and second degree, always supposing that we follow in our
reckoning with imaginary numbers the same rules that we do in
reckoning with real numbers, we always arrive again at real or
imaginary numbers, excepting when we join together a real and an
imaginary number by addition or its inverse operations. In this
case *we reach the symbol a + i . b*, where *a* and *b* stand for real numbers. Agreeably to the principle of no exception we are permitted
to reckon with *a + ib* according to the same rules of computation as
with symbols previously defined, if for the second power of *i* we
always substitute minus 1.

In the numerical combination *a + ib*, which we also call number, we have found the most general numerical form to which the
laws of arithmetic can lead, even though we wished to extend the
limits of arithmetic still further. Of course, we must represent to
ourselves here by *a* and *b* either zero or positive or negative rational
or irrational numbers. If *b* is zero, *a + ib* represents all real numbers; if *a* is zero, it stands for all purely imaginary numbers. This
general number *a + ib* is called a *complex number*, so that the complex number includes in itself as special cases all numbers heretofore defined. By the introduction of irrational, purely imaginary,
and the still more general complex numbers, all combinations become invested with meaning which the operations of the third degree can produce. For example, the fifth root of 5 is an irrational
number, the logarithm of 2 to the base 10 is an irrational number.
The logarithm of minus 1 to the base 2 is a purely imaginary number; the fourth root of minus 1 is a complex number. Indeed, we
may recognise, proceeding still further, *that every combination of two
complex numbers, by means of any of the operations of the first, second,
or third degree will lead in turn to a complex number*, that is to say,
never furnishes occasion, by application of the principle of no exception, for inventing new forms of numbers.

A certain limit is thus reached in the construction of arithmetic.
But such a limit was also twice previously reached. After the investigation of addition and its inverse operations, we reached no
other numbers except zero and positive and negative whole numbers, and every combination of such numbers by operations of the

first degree led to no new numbers. After the investigation of multiplication and its inverse operations, the positive or negative fractional numbers and "infinitely great" were added, and again we could say that the combination of two already defined numbers by operations of the first and second degree in turn also always led to numbers already defined. Now we have reached a point at which we can say that the combination of two complex numbers by all operations of the first, second, and third degree must again always lead to complex numbers; only that now such a combination does not necessarily always lead to a single number, but may lead to many regularly arranged numbers. For example, the combination "logarithm of minus one to a positive base" furnishes a countless number of results which form an arithmetical series of purely imaginary numbers. *Still, in no case now do we arrive at new classes of numbers.* But just as before the ascent from multiplication to involution brought in its train the definition of new numbers, so it is also possible that *some new operation springing out of involution as involution sprang from multiplication might furnish the germ of other new numbers which are not reducible to a + i b.* As a matter of fact, mathematicians have asked themselves this question and investigated the direct operation of the fourth degree, together with its inverse processes. The result of their investigations was, that an operation which springs from involution as involution sprang from multiplication is incapable of performing any real mathematical service; the reason of which is, that in involution the laws of commutation and association do not hold. It also further appeared that the operations of the fourth degree could not give rise to new numbers. No more so can operations of still higher degrees. With respect to quaternions, which many might be disposed to regard as new numbers, it will be evident that though quaternions are valuable means of investigation in geometry and mechanics they are not numbers of arithmetic, because the rules of arithmetic are not unconditionally applicable to them.

The building up of arithmetic is thus completed. The extensions of the domain of number are ended. It only remains to be asked why the science of arithmetic appears in its structure so logical, natural, and unarbitrary; why zero, negative, and fractional numbers appear as much derived and as little original as irrational, imaginary, and complex numbers? We answer, wholly and alone in virtue of the logical application of the monistic principle of arithmetic, the principle of no exception.

# ON THE NATURE OF MATHEMATICAL KNOWLEDGE.

"MATHEMATICALLY certain and unequivocal" is a phrase which is often heard in the sciences and in common life, to express the idea that the seal of truth is more deeply imprinted upon a proposition than is the case with ordinary acts of knowledge. We propose to investigate in this article the extent to which mathematical knowledge really is more certain and unequivocal than other knowledge.

The intrinsic character of mathematical research and knowledge is based essentially on three properties: first, on its conservative attitude towards the old truths and discoveries of mathematics; secondly, on its progressive mode of development, due to the incessant acquisition of new knowledge on the basis of the old; and thirdly, on its self-sufficiency and its consequent absolute independence.

That mathematics is the most conservative of all the sciences is apparent from the incontestability of its propositions. This last character bestows on mathematics the enviable superiority that no new development can undo the work of previous developments or substitute new in the place of old results. The discoveries that Pythagoras, Archimedes, and Apollonius made are as valid to-day as they were two thousand years ago. This is a trait which no other science possesses. The notions of previous centuries regarding the nature of heat have been disproved. Goethe's theory of colors is now antiquated. The theory of the binary combination of salts was supplanted by the theory of substitution, and this, in its turn, has also given way to newer conceptions. Think of the pro-

found changes which the conceptions of theoretical medicine, zoölogy, botany, mineralogy, and geology have undergone.   It is the
same, too, in the other sciences.   In philology, comparative linguis-
tics, and history our ideas are quite different from what they formerly
were.

In no other science is it so indispensable a condition that whatever is asserted must be true, as it is in mathematics.   Whenever,
therefore, a controversy arises in mathematics, the issue is not
whether a thing is true or not, but whether the proof might not be
conducted more simply in some other way, or whether the proposition demonstrated is sufficiently important for the advancement of the
science as to deserve especial enunciation and emphasis, or finally,
whether the proposition is not a special case of some other and
more general truth which is just as easily discovered.

Let me recall the controversy which has been waged in this
century regarding the eleventh axiom of Euclid, that only one line
can be drawn through a point parallel to another straight line.   This
discussion impugned in no wise the truth of the proposition; for
that things are true in mathematics is so much a matter of course
that on this point it is impossible for a controversy to arise.   The
discussion merely touched the question whether the axiom was
capable of demonstration solely by means of the other propositions,
or whether it was not a special property, apprehensible only by
sense-experience, of that space of three dimensions in which the
organic world has been produced and which therefore is of all others
alone within the reach of our powers of representation.   The truth
of the last supposition affects in no respect the correctness of the
axiom but simply assigns to it, in an epistemological regard, a different status from what it would have if it were demonstrable, as
was at one time thought, without the aid of the senses, and solely by
the other propositions of mathematics.

I may recall also a second controversy which arose a few decades ago as to whether all continuous functions were differentiable.
In the outcome, continuous functions were defined that possessed
no differential coefficient, and it was thus learned that certain truths
which were enunciated unconditionally by Newton, Leibnitz, and

their mathematical successors, required qualification.   But this did not invalidate at all the correctness of the method of differentiation, nor its application in all practical cases; the theoretical speculations pursued on this subject simply clarified ideas and sifted out the conditions upon which differentiability depended.   Happily the gifted minds who invent the new methods and open up the new paths of research in mathematics, are not deterred by the fear that a subsequent generation gifted with unusual acumen will spy out isolated cases in which their methods fail.   Happily the creators of the differential calculus pushed onward without a thought that a critical posterity would discover exceptions to their results.   In every great advance that mathematics makes, the clarification and scrutinisation of the results reached are reserved necessarily for a subsequent period, but with it the demonstration of those results is more rigorously established.   Despite all this, however, in no science does cognition bear so unmistakably the imprint of truth as in pure mathematics.   And this fact bestows on mathematics its conservative character.

This conservative character again is displayed in the *objects* of mathematical research.   The physician, the historian, the geographer, and the philologist have to-day quite different fields of investigation from what they had centuries ago.   In mathematics, too, every new age gives birth to new problems, arising partly from the advance of the science itself, and partly also from the advance of civilisation, where improvements in the other sciences bring in their train new problems that are constantly taxing afresh the resources of mathematics.   But despite all this, in mathematics more than in any other science problems exist that have played a rôle for hundreds, nay, for thousands of years.

In the oldest mathematical manuscript which we possess, the Rhind Papyrus of the British Museum, which dates back to the eighteenth century before Christ, and whose decipherment we owe to the industry of Eisenlohr, we find an attempt to solve the problem of converting a circle into a square of equal area, a problem whose history covers a period of three and a half thousand years. For it was not until 1882 that a rigorous proof was given of the impossibility of solving this problem exactly by the use of straight edge and compasses alone. (See pp. 116, 141–143.)

It is, of course, the insoluble problems that have the longest history; partly because it is harder to show that a thing is impossible than that it is possible, and, on the other hand, because problems that have long defied solution are ever evoking anew the spirit of inquiry and the ambition of mathematicians, and because the uncertainty of insolubility lends to such problems a peculiar charm. Of the geometrical problems that have occupied competent and incompetent minds from the time of the ancient Greeks to the present may be mentioned in addition to the squaring of the circle two others that are also perhaps well-known to educated readers, at least by name : the trisection of the angle and the Delic problem of the duplication of the cube. All three problems involve the condition, which is often overlooked by lay readers, that only straight edge and compasses shall be employed in the constructions. In the trisection of the angle any angle is assigned, and it is required to find the two straight lines which divide the angle into three equal parts. In the Delic problem the edge of a cube is given and the edge of a second cube is sought, containing twice the volume of the first cube. In Greece, in the golden age of the sciences, when all scholars had to understand mathematics, it was a fashionable requisite almost to have employed oneself on these famous problems.

Fortunately for us, these problems were insoluble. For in their ambition to conquer them it came to pass that men busied themselves more and more with geometry, and in this way kept constantly discovering new truths and developing new theories, all of which perhaps might never have been done if the problems had been soluble and had early received their solutions. Thus is the struggle after truth often more fruitful than the actual discovery of truth. So, too, although in a slightly different sense, the apophthegm of Lessing is confirmed here, that the search for truth is to be preferred to its possession.

Whilst the three above-named problems are now acknowledged to be insoluble, and have ceased, therefore, to stimulate mathematical inquiry, there are of course other problems in mathematics whose solution has been sought for a long time, but not yet reached, and in the case of which there is no reason for supposing that they are insoluble. Of such problems the two following perhaps have found their way out of the isolated circles of mathematicians and have become more or less known to other scholars. I refer to the astronomical Problem of Three Bodies and to the problem of the frequency of prime numbers. The first of these two problems assumes three or more heavenly bodies whose movements are mutually influenced by one another according to Newton's law of gravitation, and requires the exact determination of the path which each body describes. The second problem requires the construction of a formula which shall tell how many prime numbers there are below a certain given number. So far these two problems have been solved only approximately, and not with absolute mathematical exactness.

If the eternal and inviolable correctness of its truths lends to mathematical research, and therefore also to mathematical knowledge, a *conservative* character, on the other hand, by the continuous outgrowth of new truths and methods from the old, *progressiveness* is also one of its characteristics. In marvellous profusion old knowledge is augmented by new, which has the old as its necessary condition, and, therefore, could not have arisen had not the old preceded it. The indestructibility of the edifice of mathematics renders it possible that the work can be carried to ever loftier and loftier heights without fear that the highest stories shall be less solid and safe than the foundations, which are the axioms, or the lower stories, which are the elementary propositions. But it is necessary for this that all the stones should be *properly fitted together*; and it would be idle labor to attempt to lay a stone that belonged above in a place below. A good example of a stone of this character belonging in what is now the uppermost layer of the edifice, is Lindemann's demonstration of the insolubility of the quadrature of the circle, a demonstration of which interesting simplifications have been given by several mathematicians, including Weierstrass and Felix Klein. Lindemann's demonstration could not have been produced in the preceding century, because it rests necessarily on theories whose development falls in the present century. It is true,

Lambert succeeded in 1761 in demonstrating the irrationality of the ratio of the circumference of a circle to its diameter, or, which is the same thing, the irrationality of the ratio of the area of a circle to the area of the square on its radius. Afterwards, Lambert also supplied a proof that it was impossible for this ratio to be the square root of a rational number. But this was the first step only in a long journey. The attempt to prove that the old problem is insoluble was still destined to fail. An astounding mass of mathematical investigations were necessary before the demonstration could be successfully accomplished.

As we see, the majority of the mathematical truths now possessed by us presuppose the intellectual toil of many centuries. A mathematician, therefore, who wishes to-day to acquire a thorough understanding of modern research in this department, must think over again in quickened tempo the mathematical labors of several centuries. This constant dependence of new results on old ones stamps mathematics as a science of uncommon exclusiveness and renders it generally impossible to lay open to uninitiated readers a speedy path to the apprehension of the higher mathematical truths. For this reason, too, the theories and results of mathematics are rarely adapted for popular presentation. There is no royal road to the knowledge of mathematics, as Euclid once said to the first Egyptian Ptolemy. This same inaccessibility of mathematics, although it secures for it a lofty and aristocratic place among the sciences, also renders it odious to those who have never learned it, and who dread the great labor involved in acquiring an understanding of the questions of modern mathematics. Neither in the languages nor in the natural sciences are the investigations and results so closely interdependent as to make it impossible to acquaint the un-initiated student with single branches or with particular results of these sciences, without causing him to go through a long course of preliminary study.

The third trait which distinguishes mathematical research is its self-sufficiency. In philology the field of inquiry is the organic one of languages, and philology, therefore, is dependent in its investigations on the mode of development of languages, which is more or less accidental. Its task is connected with something which is given to it from without and which it cannot alter. It is much the same with the science of history, which must contemplate the history of mankind as it has actually occurred. Also zoölogy, botany, mineralogy, geology, and chemistry work with given data. In order not to become involved in futile speculations the last-mentioned sciences are constantly and inevitably obliged to revert to observations by the senses. It is then their task to link together these individual observations by bonds of causality and in this way to erect from the single stones an edifice, the view of which will render it easier for limited human intelligence to comprehend nature. Physics of all sciences stands nearest to mathematics in this respect, because unlike the other sciences she is generally in need of only a few observations in order to proceed deductively. But physics, too, must resort to observations of nature, and could not do without them for any length of time.

Mathematics alone, after certain premises have been permanently established, is able to pursue its course of development independently and unmindful of things outside of it. It can leave entirely unnoticed, questions and influences emanating from the outer world, and continue nevertheless its development.

As regards geometry, the first beginnings of this science did indeed take their origin in the requirements of practical life. But it was not long before it freed itself from the restrictions of the practical art to which it owed its birth. Herodotus recounts that geometry had its origin in Egypt where the inundations of the Nile obliterated the boundaries of the riparian estates, and by giving rise to frequent disputes constantly compelled the inhabitants to compare the areas of fields of different shapes. But with the early Greek mathematicians, who were the heirs of the Egyptian art of measurement, geometry appeared as a science which men pursued for its own sake without a thought of how their intellectual discoveries could be turned to practical account.

Nevertheless, although the workers in the domain of pure mathematics are not stimulated by the thought that their researches are likely to be of practical value, yet that result is still frequently realised, often after the lapse of centuries. The history of mathematics shows numerous instances of mathematical results which were originally the outcome of a mere desire to extend the science, suddenly receiving in astronomy, mechanics, or in physics practical applications which their originators could scarce have dreamt of. Thus Apollonius erected in ancient times the stately edifice of the properties of conic sections, without having any idea that the planets moved about the sun in conic sections, and that a Kepler and a Newton were one day to come who should apply these properties to explaining and calculating the motions of the planets about the sun. The question of the practical availability of its results in other fields has at no period exercised more than a subordinate influence on mathematical inquiry. Particularly is this true of *modern* mathematical research, whether the same consist in the extended development of isolated theories or in uniting under a higher point of view theories heretofore regarded as different.*

This independence of its character has rendered the results of pure mathematics independent also of the accidental direction which the development of civilisation has taken on our planet; so that the remark is not altogether without justification, that if beings endowed with intelligence existed on other planets, the truths of mathematics would afford the only basis of an understanding with them. Uninterruptedly and wholly from its own resources mathematics has built itself up. It is scarcely credible to a person not versed in the science, that mathematicians can derive satisfaction from the comfortless and wearisome operation of heaping up demonstration on demonstration, of rivetting truth on truth, and of tormenting themselves with self-imposed problems, whose solution stands no one in stead, and affords satisfaction to no one but the solver himself. Yet this self-sufficiency of mathematicians becomes a little more intelligible when we reflect that the progress which has been made, particularly in the last few decades, and which is uninfluenced from without, does not consist solely in the accumulation of new truths

*Cf. Felix Klein, "Remarks Given at the Opening of the Mathematical and Astronomical Congress at Chicago." *The Monist* (Vol. IV, No. 1, October, 1893).

and in the enunciation of new problems, nor solely in deductions
and solutions, but culminates rather in the discovery of new methods and points of view in which the old disconnected and isolated
results appear suddenly in a new connexion or as different interpre-
tations of a common fundamental truth, or finally, as a single organic whole.

Thus, for example, the idea of representing imaginary and complex numbers in a plane, two apparently different branches, the theory
of dividing the circumference of a circle into any given number of
equal parts, and the theory of the solutions of the equation $x^n = 1$,
have been made to exhibit an extremely simple connexion with one
another which enables us to express many a truth of algebra in a
corresponding truth of geometry and *vice versa*.    Another example is
afforded by the discovery which we chiefly owe to Alfred Clebsch,
of the relation which subsists between the higher theory of functions and the theory of algebraic curves, a relation which led to the
discovery of the condition under which two curves can be co-ordi-
nated to each other, point for point, and hence also adequately represented on each other.    Of course such combinations and extensions of view possess a much greater charm for the mathematician
than the mere accumulation of truths and solutions, whose fascina-
tion consists entirely in their truth or correctness.

From these three cardinal characteristics, now, which distinguish mathematical *research* from research in other fields, we may
gather at once the three positive characteristics that distinguish
mathematical *knowledge* from other knowledge.    They may be briefly
expressed as follows; first, mathematical knowledge bears more
distinctly the imprint of truth on all its results than any other
kind of knowledge ; secondly, it is always a sure preliminary step to
the attainment of other correct knowledge ;   thirdly, it has no need
of other knowledge.    Naturally, however, there are associated with
these characteristics which place mathematical knowledge high
above all other knowledge, other characteristics which somewhat
counterbalance the great superiority which mathematics thus appears to have over the other sciences.    In order to show more distinctly the nature of these characteristics, which we prefer to call

negative, we shall select and confine our remarks to a branch which is commonly taken to be synonymous with mathematics, namely, to arithmetic in the broadest sense of the word.

The subject of inquiry in arithmetic is numbers and their combinations. On this account arithmetic is, of all sciences, most free from what lies outside its boundaries. Perception by the senses is necessary only in an extremely insignificant measure for the understanding of its definitions and premises. It is possible to acquaint a person who lacks both sight and hearing with the fundamental principles of arithmetic solely by the medium of "time." Such a person needs only the sense of feeling. By slight excitations of his skin, induced at equal or unequal intervals of time, he can be led to the notion of differences of time and hence also to the notion of differences of number. Uninfluenced by matter and force, independently, too, of the properties of geometrical magnitudes, arithmetic could be conducted solely by its own intrinsic potencies to its highest goals, drawing deductively truth from truth, without a break.

But what sort of a science should we arrive at by this method of procedure? Nothing but a gigantic web of self-evident truths. For, once we admit the first notions and premises to which a man thus bereft of his senses can be led, we are compelled of necessity also to admit the derivative results of arithmetic. If the beginnings of arithmetic appear self-evident, the rest of it, too, bears this character. Owing to this deductive character of arithmetic, and to its exemption from influence from without, this science appears to one person extremely attractive, while to another it appears extremely repulsive, according as each is constituted. Be that as it may, however, a finished and complete science of this character subserves no purpose in the comprehension of the world, or in the advancement of civilisation. Hence, an arithmetic which heaps up theorem on theorem with never a thought of how its results are to be turned to practical account in the acquisition of knowledge in other fields, resembles an inquisitive physician, who, taking up his abode in a desert, should arrive there at momentous results in bacteriology, but should bear them with him to his grave, without their ever redounding to the benefit of humanity. The value of arithmetical knowledge lies entirely in its applications.   But this constitutes no reason why many mathematicians, pursuing their purely deductive bent of mind, should not devote themselves exclusively to pure arithmetical developments and leave it to others at the proper time to turn to the material profit of the world the capital which they have garnered.

Geometry, on the other hand, must have recourse in a much higher degree than arithmetic to the outside world for its first notions and premises.   The axioms of geometry are nothing but facts of experience perceived by our senses.   The geometry which Bolyai, Lobachévski,  Gauss, Riemann, and Helmholtz created and which is both independent of the eleventh axiom of Euclid and perfectly free from self-contradictions, has supplied an epistemological demonstration that geometry is a science that rests on the observation of nature, and therefore in the correct sense of the word, is a natural science.

Yet what a difference there is, for instance, between geometry and chemistry!  Both derive their constructive materials from sense-perception.   But whilst geometry is compelled to draw only its first results from observation and is then in a position to move forward deductively to other results without being under the necessity of making fresh observations, chemistry on the other hand is still compelled to make observations and to have recourse to nature.

It follows, therefore, that a given act of geometrical knowledge and a given act of chemical knowledge are with respect to the certainty of the truth they contain not qualitatively but only quantitatively different.   In chemistry the probability of error is greater than in geometry, because more numerous and more difficult observations have to be made there than in geometry, where only the very first premises, which no man with sound senses could ever impugn, rest on observation.

The preceding reflexions deprive mathematical knowledge of that degree of certainty and incontestability which is commonly attributed to it when we say a thing is "mathematically certain." This certainty is lessened still more as we pass to the semi-mathematical sciences, where mechanics has the first claim to our attention. All the notions of mechanics, and consequently of all the other departments of physics, are composed, by multiplication or division, of three fundamental notions—length, time, and mass. That is to say, to the notions of geometry resting on length and its powers, two other fundamental notions, time and mass, are added, which, joined to that of length, lead to the notions of force, work, horse-power, atmospheric pressure, etc. The knowledge of mechanics, thus, highly certain though it be, is rendered less certain than that of geometry and *a fortiori* than that of arithmetic. The uncertainty of knowledge continues to increase in branches which are still more remote from mathematics, owing to the increasing complexity of the observational material which must here be put to the test.

Still, although mathematical knowledge does not lead to absolutely certain results, it yet invests known results with incomparably greater trustworthiness than does the knowledge of the other sciences. But after all, it remains a useless accumulation of capital so long as it is not turned to practical account in other sciences, such as metaphysics, physics, chemistry, biology, political economy, etc. Hence also arises an obligation on the part of the other sciences, so to shape their problems and investigations that they can be made susceptible of mathematical treatment. Then will mathematics gladly perform her duty. The moment a science has advanced far enough to permit of the mathematical formulation of its problems, mathematics will not be slow to treat and to solve these problems. Mathematical knowledge, aristocratic as it may appear by the greater certainty of its results, will, so far as the advancement of human kind is concerned, never be more than a useless mass of self-evident truths, unless it constantly places itself in the service of the other sciences.

# THE MAGIC SQUARE.

## I.

### INTRODUCTORY.

AMONG the philosophies of modern times there is none which has emphasised so much the importance of form and formal thought as the monism of *The Monist*. An expression of this philosophy is found in the following passages :

"The order that prevails among the facts of reality is due to the laws of form. Upon the order of the world depends its cognisability.

".... The laws of form are no less eternal than are matter and energy and 'Verily I say unto you, till heaven and earth pass, one jot or one tittle shall in no wise pass from the law !'

"The laws of form and their origin have been a puzzle to all philosophers. 'Ay, there's the rub !' The difficulties of Hume's problem of causation, of Kant's *a priori*, of Plato's ideas, of Mill's method of deduction, etc., etc., all arise from a one-sided view of form and the laws of form and formal thought."

Considering the great results which engineering and other applied sciences accomplish through the assistance of mathematics, we must confess that the forms of thought are wonderful indeed, and it is not at all astonishing that the primitive thinkers of mankind when the importance of the laws of formal thought in some way or another first dawned on their minds, attributed magic powers to numbers and geometrical figures.

We shall devote the following pages to a brief review of magic squares, the consideration of which has made many a man believe in mysticism. And yet there is no mysticism about them unless we either consider everything mystical, even that twice two is four, or join the sceptic in his exclamation that we can truly not

ALBERT DÜRER'S ENGRAVING

# MELANCHOLY

OR THE

GENIUS OF THE INDUSTRIAL SCIENCE OF MECHANICS.

know whether twice two might not be five in other spheres of the universe.

The author of the short article on "Magic Squares" in the English Cyclopædia (Vol. III, p. 415), presumably Prof. De Morgan, says:

"Though the question of magic squares be in itself of no use, yet it belongs to a class of problems which call into action a beneficial species of investigation. Without laying down any rules for their construction, we shall content ourselves with destroying their magic quality, and showing that the non-existence of such squares would be much more surprising than their existence."

This is the point. There obtains a symphonic harmony in mathematics which is the more startling the more obvious and self-evident it appears to him who understands the laws that produce this symphonic harmony.

\* \* \*

On the wood-cut named "Melancholia"\* of the famous Nuremberg painter, Albrecht Dürer, is found among a number of other

---

\* The term melancholy meant in Dürer's time, as it did also in Shakespeare's and Milton's, "thought or thoughtfulness." Says Milton in *Il Penseroso:*

" Hail, thou Goddess, sage and holy,
Hail divinest melancholy
Whose saintly visage is too bright
To hit the sense of human sight,
And therefore to our weaker view
O'erlaid with black, staid Wisdom's hue.—I, 12.

Thought that does not lead to action produces a gloomy state of mind. Thoughtfulness which cannot find a way out of itself is that melancholy which engenders weakness,—a truth which is illustrated in Hamlet. Shakespeare still uses the words thought and melancholy as synonyms, saying:

" The native hue of resolution
Is sicklied o'er with the pale cast of thought."

Dürer's melancholy does not represent the gloominess of thought, but the power of invention. Soberness and even a certain sadness are considered only as an element of this melancholy, but on the whole the genius of thought appears bright, self-possessed, and strong.

Dürer represents the Science of Mechanical Invention as a winged female figure musing over some problem. Scattered on the floor around her lie some of the simple tools used in the sixteenth century. A ladder leans against the house, that assists in climbing otherwise inaccessible heights. A scale, an hour-glass, a bell, and the magic square are hanging on the wall behind her.

At a distance a bat-like creature, being the gloom of melancholy, hovers in the air like a dark cloud, but the sun rises above the horizon, and at the happy middle between these two extremes stands the rainbow of serene hope and cheerful confidence.

emblems, which the reader will notice in our reproduction of the cut, the subjoined square.  This arrangement of the sixteen natural num-

| 1 | 14 | 15 | 4 |
|---|----|----|---|
| 12 | 7 | 6 | 9 |
| 8 | 11 | 10 | 5 |
| 13 | 2 | 3 | 16 |

Fig. 1.

bers from 1 to 16 possesses the remarkable property that the same sum 34 will always be obtained whether we add together the four figures of any of the horizontal rows or the four of any vertical row or the four which lie in either of the two diagonals.  Such an arrangement of numbers is termed a magic square, and the square which we have reproduced above is *the first magic square which is met with in the Christian Occident.*

Like chess and many of the problems founded on the figure of the chess-board, the problem of constructing a magic square also probably traces its origin to Indian soil.  From there the problem found its way among the Arabs, and by them it was brought to the Roman Orient.  Finally, since Albrecht Dürer's time, the scholars of Western Europe also have occupied themselves with methods for the construction of squares of this character.

The oldest and the simplest magic square consists of the quadratic arrangement of the nine numbers from 1 to 9 in such a manner that the sum of each horizontal, vertical, or diagonal row, always remains the same, namely 15.  This square is the adjoined.

| 2 | 7 | 6 |
|---|---|---|
| 9 | 5 | 1 |
| 4 | 3 | 8 |

Fig. 2.

Here, we will find, 15 always comes out whether we add 2 and 7 and 6, or 9 and 5 and 1, or 4 and 3 and 8, or 2 and 9 and 4, or 7 and 5 and 3, or 6 and 1 and 8, or 2 and 5 and 8, or 6 and 5 and 4.

The question naturally presents itself, whether this condition of the constant equality of the added sum also remains fulfilled when the numbers are assigned different places.  It may be easily shown however that 5 necessarily must occupy the middle place, and that the even numbers must stand in the corners. This being so, there are but 7 additional arrangements possible, which differ from the arrangement above given and from one another only in the respect that the rows at the top, at the left, at the bottom, and at the right, exchange places with one another and that in addition a mirror be imagined present with each arrangement. So too from Dürer's square of 4 times 4 places, by transpositions, a whole set of new correct squares may be formed. A magic square of the 4 times 4 numbers from 1 to 16 is formed in the simplest manner as follows. We inscribe the numbers from 1 to 16 in their natural order in the squares, thus:

| 1 | 2 | 3 | 4 |
|----|----|----|----|
| 5 | 6 | 7 | 8 |
| 9 | 10 | 11 | 12 |
| 13 | 14 | 15 | 16 |

Fig. 3.

We then leave the numbers in the four corner-squares, viz. 1, 4, 13, 16, as well also as the numbers in the four middle-squares, viz. 6, 7, 10, 11, in their original places; and in the place of the remaining eight numbers, we write the complements of the same with respect to 17: thus 15 instead of 2, 14 instead of 3, 12 instead of 5, 9 instead of 8, 8 instead of 9, 5 instead of 12, 3 instead of 14, and 2 instead of 15. We obtain thus the magic square

| 1 | 15 | 14 | 4 | =34 |
|----|----|----|----|----|
| 12 | 6 | 7 | 9 | =34 |
| 8 | 10 | 11 | 5 | =34 |
| 13 | 3 | 2 | 16 | =34 |
| 34 | 34 | 34 | 34 | |

Fig. 4.

from which the same sum 34 always results. It is an interesting property of this square that any four numbers which form a rectangle or square about the centre also always give the same sum 34 ; for example, 1, 4, 13, 16, or 6, 7, 10, 11, or 15, 14, 3, 2, or 12, 9, 5, 8, or 15, 8, 2, 9, or 14, 12, 3, 5. We may easily convince ourselves that this square is obtainable from the square of Dürer by interchanging with one another the two middle vertical rows.

## II.

## EARLY METHODS FOR THE CONSTRUCTION OF ODD-NUMBERED SQUARES.

Since early times rules have also been known for the construction of magic squares of more than 3 times 3, or 4 times 4 spaces. In the first place, it is easy to calculate the sum which in the case of any given number of cells must result from the addition of each row. We take the determinate number of cells in each side of the square which we have to fill, multiply that number by itself, add 1, again multiply the number thus obtained by the number of the cells in each side, and, finally, divide the product by 2. Thus, with 4 times 4 cells or squares, we get: 4 times 4 are 16, 16 and 1 are 17, and one half of 17 times 4 is 34. Similarly, with 5 times 5 squares, we get: 5 times 5 are 25, and 1 makes 26, and the half of 26 times 5 is 65. Analogously, for 6 times 6 squares the summation 111 is obtained, for 7 times 7 squares 175, for 8 times 8 squares 260, for 9 times 9 squares 369, for 10 times 10 squares 505, and so on. The Hindu rule for the construction of magic squares whose roots are odd, may be enunciated as follows: To start with, write 1 in the centre of the topmost row, then write 2 in the lowest space of the

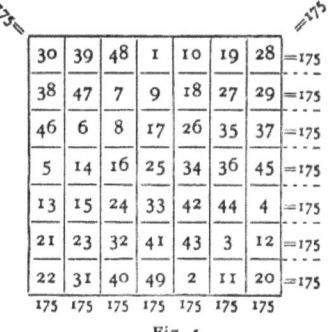

Fig. 5.

vertical column next adjacent to the right, and then so inscribe the remaining numbers in their natural order in the squares diagonally upwards towards the right, that on reaching the right-hand margin the inscription shall be continued from the left-hand margin in the row just above, and on reaching the upper margin shall be continued from the lower margin in the column next adjacent to the right, noting that whenever we are arrested in our progress by a square already occupied we are to fill out the square next beneath the one we have last filled.   In this manner, for example, the last preceding square of 7 times 7 cells is formed, in which the reader is requested to follow the numbers in their natural sequence (Fig. 5).

For the next further advancements of the theory of magic squares and of the methods for their construction we are indebted to the Byzantian Greek, Moschopulus, who lived in the fourteenth century ; also, after Albrecht Dürer who lived about the year 1500, to the celebrated arithmetician Adam Riese, and to the mathematician Michael Stifel, which two last lived about 1550.   In the seventeenth century Bachet de Méziriac, and Athanasius Kircher employed themselves on magic squares.   About 1700, finally, the French mathematicians De la Hire and Sauveur made considerable contributions to the theory.   In recent times mathematicians have concerned themselves much less about magic squares, as they have indeed about mathematical recreations generally.   But quite recently the Brunswick mathematician Scheffler has put forth his own and other's studies on this subject in an elegant form.

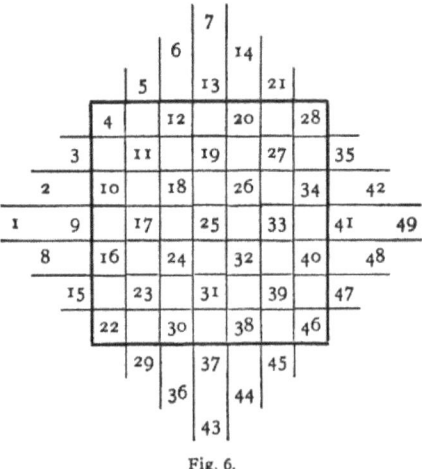

Fig. 6.

The best known of the various methods of constructing magic squares of an odd number of cells is the following. First write the numbers in diagonal succession as in the preceding diagram (Fig. 6). After 25 cells of the square of 49 cells which we have to fill out, have thus been occupied, transfer the six figures found outside each side of the square, without changing their configuration, into the empty cells of the side directly opposite. By this method, which we owe to Bachet de Méziriac, we obtain the following magic square of the numbers from 1 to 49:

| 4 | 29 | 12 | 37 | 20 | 45 | 28 |
| 35 | 11 | 36 | 19 | 44 | 27 | 3 |
| 10 | 42 | 18 | 43 | 26 | 2 | 34 |
| 41 | 17 | 49 | 25 | 1 | 33 | 9 |
| 16 | 48 | 24 | 7 | 32 | 8 | 40 |
| 47 | 23 | 6 | 31 | 14 | 39 | 15 |
| 22 | 5 | 30 | 13 | 38 | 21 | 46 |

Fig. 7.

III.

MODERN MODES OF CONSTRUCTION OF ODD-NUMBERED SQUARES.

The reader will justly ask whether there do not exist other correct magic squares which are constructed after a different method from that just given, and whether there do not exist modes of construction which will lead to all the imaginable and possible magic squares of a definite number of cells. A general mode of construction of this character was first given for odd-numbered squares by De la Hire, and recently perfected by Professor Scheffler.

To acquaint ourselves with this general method, let us select as our example a square of 5. First we form two auxiliary squares. In the first we write the numbers from 1 to 5 five times; and in the second, five times, the following multiples of five, viz.: 0, 5, 10, 15, 20. It is clear now that by adding each of the numbers of the series from 1 to 5 to each of the numbers 0, 5, 10, 15, 20, we shall get all the 25 numerals from 1 to 25. All that additionally remains to be done therefore, is, so to inscribe the numbers that by the addition of the two numbers in any two corresponding cells each combination shall come out once and only once ; and further that in each horizontal, vertical, and diagonal row in each auxiliary square each number shall once appear. Then the required sum of 65 must necessarily result in every case, because the numbers from 1 to 5 added together make 15, and the numbers 0, 5, 10, 15, 20 make 50.

We effect the required method of inscription by imagining the numbers 1, 2, 3, 4, 5 (or 0, 5, 10, 15, 20) arranged in cyclical succession, that is 1 immediately following upon 5, and, starting from any number whatsoever, by skipping each time either none or one or two or three etc. figures. Cycles are thus obtained of the first, the second, the third etc. orders ; for example 3 4 5 1 2 is a cycle of the first order, 2 4 1 3 5 is a cycle of the second order, 1 5 4 3 2 is a cycle of the fourth order, etc. The only thing then to be looked out for in the two auxiliary squares is, that the same "cycle" order be horizontally preserved in all the rows, that the same also happens for the vertical rows, but that the cycle order in the horizontal and vertical rows is different. Finally we have only additionally to take care that to the same numbers of the one auxiliary square not like numbers but *different* numbers correspond in the other auxiliary square, that is lie in similarly situated cells. The following auxiliary squares are, for example, thus possible :

| 3 | 4 | 5 | 1 | 2 |
|---|---|---|---|---|
| 5 | 1 | 2 | 3 | 4 |
| 2 | 3 | 4 | 5 | 1 |
| 4 | 5 | 1 | 2 | 3 |
| 1 | 2 | 3 | 4 | 5 |

and

| 0 | 10 | 20 | 5 | 15 |
|---|---|---|---|---|
| 5 | 15 | 0 | 10 | 20 |
| 10 | 20 | 5 | 15 | 0 |
| 15 | 0 | 10 | 20 | 5 |
| 20 | 5 | 15 | 0 | 10 |

Fig. 8.    Fig. 9.

Adding in pairs the numbers which occupy similarly situated cells, we obtain the following correct magic square :

| 3 | 14 | 25 | 6 | 17 |
|---|---|---|---|---|
| 10 | 16 | 2 | 13 | 24 |
| 12 | 23 | 9 | 20 | 1 |
| 19 | 5 | 11 | 22 | 8 |
| 21 | 7 | 18 | 4 | 15 |

Fig. 10.

THE MAGIC SQUARE.

It will be seen that we are able thus to construct a very large number of magic squares of 5 times 5 spaces by varying in every possible manner the numbers in the two auxiliary squares. Furthermore, the squares thus formed possess the additional peculiarity, that every 5 numbers which fill out two rows that are parallel to a diagonal and lie on different sides of the diagonal also give the constant sum of 65. For example: 3 and 7, 11, 20, 24; or 10, 14 and 18, 22, 1. Altogether then the sum 65 is produced out of 20 rows or pairs of rows. On this peculiarity is dependent the fact that if we imagine an unlimited number of such squares placed by the side of, above, or beneath an initial one, we shall be able to obtain as many quadratic cells as we choose, so arranged that the square composed of any 25 of these cells will form a correct magic square, as the following figure will show :

| 2  | 13 | 24 | 10 | 16 | 2  | 13 | 24 | 10 | 16 | 2  |
|----|----|----|----|----|----|----|----|----|----|----|
| 9  | 20 | 1  | 12 | 23 | 9  | 20 | 1  | 12 | 23 | 9  |
| 11 | 22 | 8  | 19 | 5  | 11 | 22 | 8  | 19 | 5  | 11 |
| 18 | 4  | 15 | 21 | 7  | 18 | 4  | 15 | 21 | 7  | 18 |
| 25 | 6  | 17 | 3  | 14 | 25 | 6  | 17 | 3  | 14 | 25 |
| 2  | 13 | 24 | 10 | 16 | 2  | 13 | 24 | 10 | 16 | 2  |
| 9  | 20 | 1  | 12 | 23 | 9  | 20 | 1  | 12 | 23 | 9  |
| 11 | 22 | 8  | 19 | 5  | 11 | 22 | 8  | 19 | 5  | 11 |
| 18 | 4  | 15 | 21 | 7  | 18 | 4  | 15 | 21 | 7  | 18 |
| 25 | 6  | 17 | 3  | 14 | 25 | 6  | 17 | 3  | 14 | 25 |
| 2  | 13 | 24 | 10 | 16 | 2  | 13 | 24 | 10 | 16 | 2  |
| 9  | 20 | 1  | 12 | 23 | 9  | 20 | 1  | 12 | 23 | 9  |
| 11 | 22 | 8  | 19 | 5  | 11 | 22 | 8  | 19 | 5  | 11 |

Fig. 11.

Every square of every 25 of these numbers, as for example the two dark-bordered ones, possesses the property that the addition of the horizontal, vertical, and diagonal rows gives each the same sum, 65.

As an example of a higher number of cells we will append here a magic square of 11 times 11 spaces formed by the general method of De la Hire from the two auxiliary squares of Figs. 12 and 13. From these two auxiliary squares we obtain by the addition of the two numbers of every two similarly situated cells, the magic square, exhibited in Diagram 14, in which each row gives the same sum 671.

| 1 | 2 | 3 | 4 | 5 | 6 | 7 | 8 | 9 | 10 | 11 |
|---|---|---|---|---|---|---|---|---|---|---|
| 3 | 4 | 5 | 6 | 7 | 8 | 9 | 10 | 11 | 1 | 2 |
| 5 | 6 | 7 | 8 | 9 | 10 | 11 | 1 | 2 | 3 | 4 |
| 7 | 8 | 9 | 10 | 11 | 1 | 2 | 3 | 4 | 5 | 6 |
| 9 | 10 | 11 | 1 | 2 | 3 | 4 | 5 | 6 | 7 | 8 |
| 11 | 1 | 2 | 3 | 4 | 5 | 6 | 7 | 8 | 9 | 10 |
| 2 | 3 | 4 | 5 | 6 | 7 | 8 | 9 | 10 | 11 | 1 |
| 4 | 5 | 6 | 7 | 8 | 9 | 10 | 11 | 1 | 2 | 3 |
| 6 | 7 | 8 | 9 | 10 | 11 | 1 | 2 | 3 | 4 | 5 |
| 8 | 9 | 10 | 11 | 1 | 2 | 3 | 4 | 5 | 6 | 7 |
| 10 | 11 | 1 | 2 | 3 | 4 | 5 | 6 | 7 | 8 | 9 |

Fig. 12.

| 0 | 11 | 22 | 33 | 44 | 55 | 66 | 77 | 88 | 99 | 110 |
|---|---|---|---|---|---|---|---|---|---|---|
| 33 | 44 | 55 | 66 | 77 | 88 | 99 | 110 | 0 | 11 | 22 |
| 66 | 77 | 88 | 99 | 110 | 0 | 11 | 22 | 33 | 44 | 55 |
| 99 | 110 | 0 | 11 | 22 | 33 | 44 | 55 | 66 | 77 | 88 |
| 11 | 22 | 33 | 44 | 55 | 66 | 77 | 88 | 99 | 110 | 0 |
| 44 | 55 | 66 | 77 | 88 | 99 | 110 | 0 | 11 | 22 | 33 |
| 77 | 88 | 99 | 110 | 0 | 11 | 22 | 33 | 44 | 55 | 66 |
| 110 | 0 | 11 | 22 | 33 | 44 | 55 | 66 | 77 | 88 | 99 |
| 22 | 33 | 44 | 55 | 66 | 77 | 88 | 99 | 110 | 0 | 11 |
| 55 | 66 | 77 | 88 | 99 | 110 | 0 | 11 | 22 | 33 | 44 |
| 88 | 99 | 110 | 0 | 11 | 22 | 33 | 44 | 55 | 66 | 77 |

Fig. 13.

| 1 | 13 | 25 | 37 | 49 | 61 | 73 | 85 | 97 | 109 | 121 |
|---|---|---|---|---|---|---|---|---|---|---|
| 36 | 48 | 60 | 72 | 84 | 96 | 108 | 120 | 11 | 12 | 24 |
| 71 | 83 | 95 | 107 | 119 | 10 | 22 | 23 | 35 | 47 | 59 |
| 106 | 118 | 9 | 21 | 33 | 34 | 46 | 58 | 70 | 82 | 94 |
| 20 | 32 | 44 | 45 | 57 | 69 | 81 | 93 | 105 | 117 | 8 |
| 55 | 56 | 68 | 80 | 92 | 104 | 116 | 7 | 19 | 31 | 43 |
| 79 | 91 | 103 | 115 | 6 | 18 | 30 | 42 | 54 | 66 | 67 |
| 114 | 5 | 17 | 29 | 41 | 53 | 65 | 77 | 78 | 90 | 102 |
| 28 | 40 | 52 | 64 | 76 | 88 | 89 | 101 | 113 | 4 | 16 |
| 63 | 75 | 87 | 99 | 100 | 112 | 3 | 15 | 27 | 39 | 51 |
| 98 | 110 | 111 | 2 | 14 | 26 | 38 | 50 | 62 | 74 | 86 |

Fig. 14

IV.

## EVEN-NUMBERED SQUARES.

Of magic squares having an even number of places we have hitherto had to deal only with the square of 4. To construct squares of this description having a higher even number of places, different and more complicated methods must be employed than for squares of odd numbers of places. However, in this case also, as in dealing with the square of 4, we start with the natural sequence of the numbers and must then find the complements of the numbers with respect to some other certain number (as 17 in the square of 4) and also effect certain exchanges of the numbers with one another. To form, for example, a magic square of 6 times 6 places, we inscribe in the 12 diagonal cells the numbers that in the natural sequence of inscription fall into these places, then in the remaining cells the complements of the numbers that belong therein with respect to 37, and finally effect the following six exchanges, viz. of the numbers 33 and 3, 25 and 7, 20 and 14, 18 and 13, 10 and 9, and 5 and 2. In this way the following magic square is obtained.

| 1 | 35 | 34 | 3 | 32 | 6 |
|---|----|----|---|----|---|
| 30 | 8 | 28 | 27 | 11 | 7 |
| 24 | 23 | 15 | 16 | 14 | 19 |
| 13 | 17 | 21 | 22 | 20 | 18 |
| 12 | 26 | 9 | 10 | 29 | 25 |
| 31 | 2 | 4 | 33 | 5 | 36 |

Fig. 15.

This square may also be constructed by the method of De la Hire, from two auxiliary squares with the numbers 1, 2, 3, 4, 5, 6 and 0, 6, 12, 18, 24, 30 respectively. In this case, however, the vertical rows of the one square and the horizontal rows of the other must each so contain two numbers three times repeated that the summation shall always remain 21 and 90 respectively. In this manner we get the magic square last given above from the two following auxiliary squares:

| 1 | 5 | 4 | 3 | 2 | 6 |
|---|---|---|---|---|---|
| 6 | 2 | 4 | 3 | 5 | 1 |
| 6 | 5 | 3 | 4 | 2 | 1 |
| 1 | 5 | 3 | 4 | 2 | 6 |
| 6 | 2 | 3 | 4 | 5 | 1 |
| 1 | 2 | 4 | 3 | 5 | 6 |

and

| 0 | 30 | 30 | 0 | 30 | 0 |
|---|----|----|---|----|---|
| 24 | 6 | 24 | 24 | 6 | 6 |
| 18 | 18 | 12 | 12 | 12 | 18 |
| 12 | 12 | 18 | 18 | 18 | 12 |
| 6 | 24 | 6 | 6 | 24 | 24 |
| 30 | 0 | 0 | 30 | 0 | 30 |

Fig. 16.                    Fig. 17.

It is to be noted in connection with this example that here also as in the case of odd-numbered squares, it is possible so to inscribe six times the numbers from 1 to 6 that each number shall appear once and only once in each horizontal, vertical, and diagonal row; for example, in the following manner:

| 1 | 2 | 3 | 4 | 5 | 6 |
|---|---|---|---|---|---|
| 2 | 4 | 6 | 1 | 3 | 5 |
| 3 | 6 | 5 | 2 | 1 | 4 |
| 5 | 3 | 1 | 6 | 4 | 2 |
| 6 | 5 | 4 | 3 | 2 | 1 |
| 4 | 1 | 2 | 5 | 6 | 3 |

Fig. 18.

But if we attempt so to insert, in a like manner, the other set of numbers 0, 6, 12, 18, 24, 30 in a second auxiliary square, that each number of the first auxiliary square shall stand once and once only in a corresponding cell with each number of the second square, all the attempts we may make to fulfil coincidently the last named condition will result in failure. It is therefore necessary to select auxiliary squares like the two given above. It is noteworthy, that the fulfilment of the second condition is impossible only in the case of the square of 6, but that in the case of the square of 4 or of the square of 8, for example, two auxiliary squares, such as the method of De la Hire requires, are possible. Thus, taking the square of 4 we get

| 1 | 2 | 3 | 4 |
|---|---|---|---|
| 4 | 3 | 2 | 1 |
| 2 | 1 | 4 | 3 |
| 3 | 4 | 1 | 2 |

and

| 0 | 4 | 8 | 12 |
|---|---|---|---|
| 8 | 12 | 0 | 4 |
| 12 | 8 | 4 | 0 |
| 4 | 0 | 12 | 8 |

Fig. 19.         Fig. 20.

The reader may form for himself the magic square which these give.

The existence of these two auxiliary squares furnishes a key to the solution of a pretty problem at cards. If we replace, namely, the numbers 1, 2, 3, 4 by the Ace, the King, the Queen, and the Knave, and the numbers 0, 4, 8, 12 by the four suits, clubs, spades, hearts, and diamonds, we shall at once perceive that it is possible, and must be so necessarily, quadratically to arrange in such a manner the four Aces, the four Kings, the Four Queens, and the four

Knaves, that in each horizontal, vertical, and diagonal row, each one of the four suits and each one of the four denominations shall appear once and once only. The auxiliary squares above given furnish the appended solution of this problem:

| CLUBS ACE | SPADES KING | HEARTS QUEEN | DIAMONDS KNAVE |
|---|---|---|---|
| HEARTS KNAVE | DIAMONDS QUEEN | CLUBS KING | SPADES ACE |
| DIAMONDS KING | HEARTS ACE | SPADES KNAVE | CLUBS QUEEN |
| SPADES QUEEN | CLUBS KNAVE | DIAMONDS ACE | HEARTS KING |

Fig. 21.

To fix the solution of the problem in the memory, observe that, starting from the several corners, each suit and each denomination must be placed in the spots of the move of a Knight. If we fix the positions of the four cards of any one row, there will be only two possibilities left of so placing the other cards that the required condition of having each suit and each denomination once and only once in each row shall be fulfilled.

Of magic squares of an even number of places we have up to this point examined only the squares of 4 and of 6. For the sake of completeness we append here one of 8 and one of 10 places. The mode of construction of these squares is similar to the method above discussed for the lower even numbers.

| 1 | 63 | 62 | 4 | 5 | 59 | 58 | 8 |
|---|---|---|---|---|---|---|---|
| 56 | 10 | 11 | 53 | 52 | 14 | 15 | 49 |
| 48 | 18 | 19 | 45 | 44 | 22 | 23 | 41 |
| 25 | 39 | 38 | 28 | 29 | 35 | 34 | 32 |
| 33 | 31 | 30 | 36 | 37 | 27 | 26 | 40 |
| 24 | 42 | 43 | 21 | 20 | 46 | 47 | 17 |
| 16 | 50 | 51 | 13 | 12 | 54 | 55 | 9 |
| 57 | 7 | 6 | 60 | 61 | 3 | 2 | 64 |

Fig. 22.

| 1 | 99 | 3 | 97 | 96 | 5 | 94 | 8 | 92 | 10 |
|---|----|---|----|----|---|----|---|----|----|
| 90 | 12 | 88 | 14 | 86 | 85 | 17 | 83 | 19 | 11 |
| 80 | 79 | 23 | 77 | 25 | 26 | 74 | 28 | 22 | 71 |
| 31 | 69 | 68 | 34 | 66 | 65 | 37 | 33 | 62 | 40 |
| 60 | 42 | 58 | 57 | 45 | 46 | 44 | 53 | 49 | 51 |
| 50 | 52 | 43 | 47 | 55 | 56 | 54 | 48 | 59 | 41 |
| 61 | 32 | 38 | 64 | 36 | 35 | 67 | 63 | 39 | 70 |
| 21 | 29 | 73 | 27 | 75 | 76 | 24 | 78 | 72 | 30 |
| 20 | 82 | 18 | 84 | 15 | 16 | 87 | 13 | 89 | 81 |
| 91 | 9 | 93 | 4 | 6 | 95 | 7 | 98 | 2 | 100 |

Fig. 23.

The magic squares of even numbers thus constructed are not the only possible ones. On the contrary, there are very many others possible, which obey different laws of formation. It has been calculated, for example, that with the square of 4 it is possible to construct 880, and with the square of 6, *several million*, different magic squares. The number of odd-numbered magic squares constructible by the method of De la Hire is also very great. With the square of 7, the possible constructions amount to 363,916,800. With the squares of higher numbers the multitude of the possibilities increases in the same enormous ratio.

v.

MAGIC SQUARES WHOSE SUMMATION GIVES THE NUMBER
OF A YEAR.

The magic squares which we have so far considered contain only the natural numbers from 1 upwards. It is possible, however, easily to deduce from a correct magic square other squares in which a different law controls the sequence of the numbers to be inscribed. Of the squares obtained in this manner, we shall devote our attention here only to such in which, although formed by the inscription of successive numbers, the sum obtained from the addition of the rows is a determinate number which we have fixed upon beforehand, as *the number of a year*. In such a case we have simply to add to the numbers of the original square a determinate number so to be calculated, that the required sum shall each time appear. If this sum is divisible by 3, magic squares will always be obtainable with 3 times 3 spaces which shall give this sum. In such a case we divide the sum required by 3 and subtract 5 from the result in order to obtain the number which we have to add to each number of the original square. If the sum desired is even but not divisible by 4, we must then subtract from it 34 and take one fourth of the result, to obtain the number which in this case is to be added in each place. If, for example, we wish to obtain the number of the year 1890 as the resulting sum of each row, we shall have to add to each of the numbers of an ordinary magic square of 4 times 4 spaces the number 464; in other words, instead of the numbers from 1 to 16 we have to insert in the squares the numbers from 465 to 480. As the number of the year 1892 is divisible by eleven, it must be possible to deduce from the magic square constructed by us at the conclusion of Section III a second magic square in which each row of 11 cells will give the number of the year 1892. To do this, we subtract from 1892 the sum of the original square, namely 671, and divide the remainder by 11, whereby we get 111 and thus perceive that the numbers from 112 to 232 are to be inscribed in the cells of

| | | | | | | | | | | | |
|---|---|---|---|---|---|---|---|---|---|---|---|
| 112 | 124 | 136 | 148 | 160 | 172 | 184 | 196 | 208 | 220 | 232 | = 1892 |
| 147 | 159 | 171 | 183 | 195 | 207 | 219 | 231 | 122 | 123 | 135 | = 1892 |
| 182 | 194 | 206 | 218 | 230 | 121 | 133 | 134 | 146 | 158 | 170 | = 1892 |
| 217 | 229 | 120 | 132 | 144 | 145 | 157 | 169 | 181 | 193 | 205 | = 1892 |
| 131 | 143 | 155 | 156 | 168 | 180 | 192 | 204 | 216 | 228 | 110 | = 1892 |
| 166 | 167 | 179 | 191 | 203 | 215 | 227 | 118 | 130 | 142 | 154 | = 1892 |
| 190 | 202 | 214 | 226 | 117 | 129 | 141 | 153 | 165 | 177 | 178 | = 1892 |
| 225 | 116 | 128 | 140 | 152 | 164 | 176 | 188 | 189 | 201 | 213 | = 1892 |
| 137 | 151 | 163 | 175 | 187 | 199 | 200 | 212 | 224 | 115 | 127 | = 1892 |
| 174 | 186 | 198 | 210 | 211 | 123 | 114 | 126 | 138 | 150 | 162 | = 1892 |
| 200 | 221 | 222 | 113 | 125 | 137 | 149 | 161 | 173 | 185 | 107 | = 1892 |

1892 1892 1892 1892 1892 1892 1892 1892 1892 1892 1892

Fig. 24.

the square required. We get in this way the preceding square, from which *one and the same sum, namely 1892, can be obtained 44 times,* first from each of the 11 horizontal rows, secondly from each of the 11 vertical rows, thirdly from each of the two diagonal rows, and fourthly twenty additional times from each and every pair of any two rows that lie parallel to a diagonal, have together 11 cells, and lie on different sides of the diagonal, as for example, 196, 122, 158, 205, 131, 167, 214, 140, 187, 223, 149.

VI.

CONCENTRIC MAGIC SQUARES.

The acuteness of mathematicians has also discovered magic squares which possess the peculiar property that if one row after another be taken away from each side, the smaller inner squares remaining will still be magical squares, that is to say, all their rows when added will give the same sum. It will be sufficient to give two examples here of such squares, (the laws for their construction being somewhat more complicated,) of which the first has 7 times 7 and the second 8 times 8 places. The numbers within each of the dark-bordered frames form with respect to the centre smaller squares which in their own turn are magical.

| 4 | 5 | 6 | 43 | 39 | 38 | 40 |
|---|---|---|----|----|----|----|
| 49 | 15 | 16 | 33 | 30 | 31 | 1 |
| 48 | 37 | 22 | 27 | 26 | 13 | 2 |
| 47 | 36 | 29 | 25 | 21 | 14 | 3 |
| 8 | 18 | 24 | 23 | 28 | 32 | 42 |
| 9 | 19 | 34 | 17 | 20 | 35 | 41 |
| 10 | 45 | 44 | 7 | 11 | 12 | 46 |

Fig. 25.

| 1 | 56 | 55 | 11 | 53 | 13 | 14 | 57 |
|---|----|----|----|----|----|----|----|
| 63 | 15 | 47 | 22 | 42 | 24 | 45 | 2 |
| 62 | 49 | 25 | 40 | 34 | 31 | 16 | 3 |
| 4 | 48 | 28 | 37 | 35 | 30 | 17 | 61 |
| 5 | 44 | 39 | 26 | 32 | 33 | 21 | 60 |
| 59 | 19 | 38 | 27 | 29 | 36 | 46 | 6 |
| 58 | 20 | 18 | 43 | 23 | 41 | 50 | 7 |
| 8 | 9 | 10 | 54 | 12 | 52 | 51 | 64 |

Fig. 26.

In the first of these two squares the internal square of 3 times 3 places contains the numbers from 21 to 29 in such a manner that each row gives when added the sum of 75. This square lies within a larger one of 5 times 5 spaces, which contains the numbers from 13 to 37 in such a manner that each row gives the sum of 125. Finally, this last square forms part of a square of 7 times 7 places which contains the numbers from 1 to 49 so that each row gives the sum of 175.

In the second square the inner central square of 4 times 4 places contains the numbers from 25 to 40 in such a manner that each row gives the sum of 130.　This square is the middle of a square of 6 times 6 places which so contains the numbers from 15 to 50 that each row gives the sum 195.　Finally, this last square is again the middle of an ordinary magic square composed of the numbers from 1 to 64.

VII.

MAGICAL SQUARES WITH MAGICAL PARTS.

If we divide a square of 8 times 8 places by means of the two middle lines parallel to its sides into 4 parts containing each 4 times 4 spaces, we may propound the problem of so inserting the numbers from 1 to 64 in these spaces that not only the whole shall form a magic square, but also that each of the 4 parts individually shall be magical, that is to say, give the same sum for each row.　This problem also has been successfully solved, as the following diagram will show.

| 1 | 4 | 63 | 62 | 5 | 8 | 59 | 58 |
|---|---|---|---|---|---|---|---|
| 64 | 61 | 2 | 3 | 60 | 57 | 6 | 7 |
| 42 | 43 | 24 | 21 | 34 | 35 | 32 | 29 |
| 23 | 22 | 41 | 44 | 31 | 30 | 33 | 36 |
| 13 | 16 | 51 | 50 | 9 | 12 | 55 | 54 |
| 52 | 49 | 14 | 15 | 56 | 53 | 10 | 11 |
| 38 | 39 | 28 | 25 | 46 | 47 | 20 | 17 |
| 27 | 26 | 37 | 40 | 19 | 18 | 45 | 48 |

Fig. 27.

The 4 numbers in each row of any one of the sub-squares here, gives 130; so that the sum of each one of the rows of the large square will be 260.

Finally, in further illustration of this idea, we will submit to the consideration of our readers a very remarkable square of the numbers from 1 to 81.　This square, which will be found on the following page (Fig. 28), is divided by parallel lines into 9 parts, of which each contains 9 consecutive numbers that severally make up a magic square by themselves.

Wonderful as the properties of this square may appear, the law by which the author constructed it is equally simple.　We have simply to regard the 9 parts as the 9 cells of a magic square of the numbers from I to IX, and then to inscribe by the magic prescript in the square designated as I the numbers from 1 to 9, in the square

| 31 | 36 | 29 | 76 | 81 | 74 | 13 | 18 | 11 |
|----|----|----|----|----|----|----|----|----|
| 30 | 32 | 34 | 75 | 77 | 79 | 12 | 14 | 16 |
| 35 | 28 | 33 | 80 | 73 | 78 | 17 | 10 | 15 |
| 22 | 27 | 20 | 40 | 45 | 38 | 58 | 63 | 56 |
| 21 | 23 | 25 | 39 | 41 | 43 | 57 | 59 | 61 |
| 26 | 19 | 24 | 44 | 37 | 42 | 62 | 55 | 60 |
| 67 | 72 | 65 | 4 | 9 | 2 | 49 | 54 | 47 |
| 66 | 68 | 70 | 3 | 5 | 7 | 48 | 50 | 52 |
| 71 | 64 | 69 | 8 | 1 | 6 | 53 | 46 | 51 |

Fig. 28.

designated as II the numbers from 10 to 18, and so on. In this way the square above given is obtained from the following base-square:

| IV | IX | II |
|----|----|----|
| III | V | VII |
| VIII | I | VI |

Fig. 29.

VIII.

MAGIC SQUARES THAT INVOLVE THE MOVE OF THE CHESS-KNIGHT.

What one of our readers does not know the problems contained in the recreation columns of our magazines, the requirements of which are to compose into a verse 8 times 8 quadratically arranged syllables, of which every two successive syllables stand on spots so situated with respect to each other that a chess-knight can move from the one to the other? If we replace in such an arrangement the 64 successive syllables by the 64 numbers from 1 to 64, we shall obtain a knight-problem made up of numbers. Methods also exist indeed for the construction of such dispositions of numbers, which then form the foundation of the construction of the problems in the newspapers. But the majority of knight-problems of this class are the outcome of experiment rather than the product of method-

ical creation.    If however it is a severe test of patience to form a
knight-problem by experiment, it stands to reason that it is a still
severer trial to effect at the same time the additional result that the
64 numbers which form the knight-problem shall also form a magic
square.

This trial of endurance was undertaken several decades ago, by
a pensioned Moravian officer named Wenzelides, who was spending
the last days of his life in the country.  After a series of trials which
lasted years he finally succeeded in so inscribing in the 64 squares
of the chess-board the numbers from 1 to 64 that successive numbers, as well also as the numbers 64 and 1, were always removed
from one another in distance and direction by the move of a knight,
and that in addition thereto the summation of the horizontal and the
vertical rows always gave the same sum 260.   Ultimately he discovered several squares of this description, which were published in
the *Berlin Chess Journal.*   One of these is here appended :

| 47 | 10 | 23 | 64 | 49 | 2  | 59 | 6  |
|----|----|----|----|----|----|----|----|
| 22 | 63 | 48 | 9  | 60 | 5  | 50 | 3  |
| 11 | 46 | 61 | 24 | 1  | 52 | 7  | 58 |
| 62 | 21 | 12 | 45 | 8  | 57 | 4  | 51 |
| 19 | 36 | 25 | 40 | 13 | 44 | 53 | 30 |
| 26 | 39 | 20 | 33 | 56 | 29 | 14 | 43 |
| 35 | 18 | 37 | 28 | 41 | 16 | 31 | 54 |
| 38 | 27 | 34 | 17 | 32 | 55 | 42 | 15 |

Fig. 30.

The move of the knight and the equality of the summation of
the horizontal and vertical rows, therefore, are the facts to be noted
here.   The diagonal rows do *not* give the sum 260.   Perhaps some
one among our readers who possesses the time and patience will be
tempted to outdo Wenzelides, and to devise a numeral knight-problem of this kind which will give 260 not only in the horizontal and
vertical but also in the two diagonal rows.

IX.

MAGICAL POLYGONS.

So far we have considered only such extensions of the idea underlying the construction of the magic square in which the figure of the square was retained. We may however contrive extensions of the idea in which instead of a square, a rectangle, a triangle, or a pentagon, and the like, appear. Without entering into the consideration of the methods for the construction of such figures, we will give here of magical polygons simply a few examples, all supplied by Professor Scheffler:

1) The numbers from 1 to 32 admit of being written in a rectangle of 4 × 8 in such a manner that the long horizontal rows give the sum of 132 and the short vertical rows the sum of 66; thus:

| 1 | 10 | 11 | 29 | 28 | 19 | 18 | 16 |
|---|----|----|----|----|----|----|----|
| 9 | 2 | 30 | 12 | 20 | 27 | 7 | 25 |
| 24 | 31 | 3 | 21 | 13 | 6 | 26 | 8 |
| 32 | 23 | 22 | 4 | 5 | 14 | 15 | 17 |

Fig. 31.

2) The numbers from 1 to 27 admit of being so arranged in three regular triangles about a point which forms a common centre, that each side of the outermost triangle will present 6 numbers of the total summation 96 and each side of the middle triangle 4 numbers whose sum is 61; as the following figure shows:

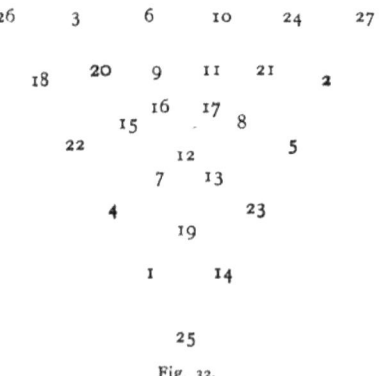

Fig. 32.

3) The numbers from 1 to 80 admit of being formed about a
point as common centre into 4 pentagons, such that each side of
the first pentagon from within contains two numbers, each side of
the second pentagon four numbers, each of the third six numbers,
and each side of the fourth, outermost pentagon eight numbers.
The sum of the numbers of each side of the second pentagon is 122,
the sum of those of each side of the third pentagon is 248, and that
of those of each side of the fourth pentagon 254. Furthermore, the
sum of any four corner numbers lying in the same straight line with
the centre, is also the same; namely, 92.

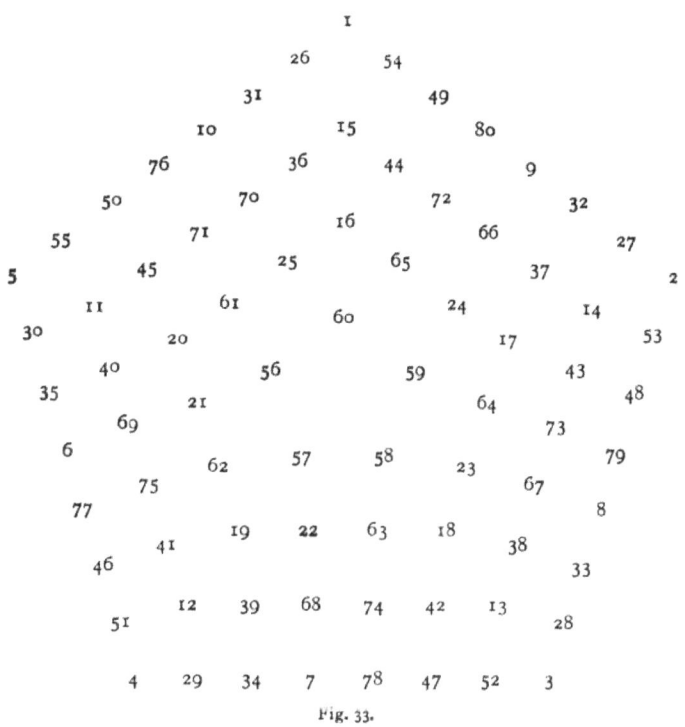

Fig. 33.

4) The numbers from 1 to 73 admit of being arranged about a
centre, in which the number 37 is written, into three hexagons which
contain respectively 3, 5, and 7 numbers in each side and possess
the following pretty properties. Each hexagon always gives the
same sum, not only when the summation is made along its six sides,
but also when it is made along the six diameters that join its corners

and along the six that are constructed at right angles to its sides ; this sum, for the first hexagon from within, is 111, for the second 185, and for the third 259.

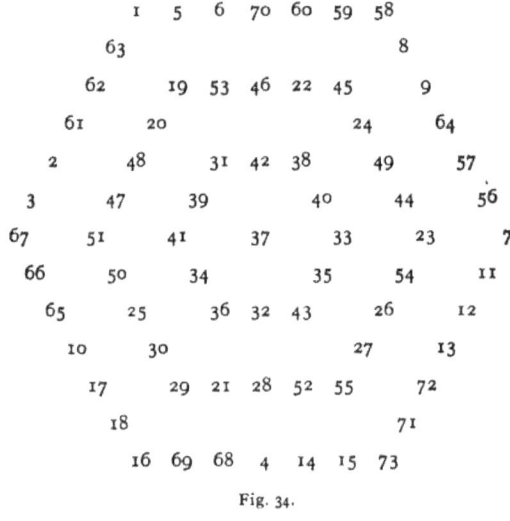

Fig. 34.

X.

MAGIC CUBES.

Several inquirers, particularly Kochansky (1686), Sauveur (1710), Hugel (1859), and Scheffler (1882), have extended the principle of the magic squares of the plane to three-dimensioned space. Imagine a cube divided by planes parallel to its sides and equidistant from one another, into cubical compartments. The problem is then, so to insert in these compartments the successive natural numbers that every row from the right to the left, every row from the front to the back, every row from the top to the bottom, every diagonal of a square, and every principal diagonal passing through the centre of the cube shall contain numbers whose sum is always the same. For 3 times 3 times 3 compartments, a magic cube of this description is not constructible. For 4 times, 4 times 4 compartments a cube is constructible such that any row parallel to an edge of the cube and every principal diagonal give the sum of 130. To obtain a magic cube of 64 compartments, imagine the numbers which belong in the compartments written on the upper surface of the same and the numbers then taken off in layers of 16 from the top downwards. We obtain thus 4 squares of 16 cells each, which together make up the magic cube; as the following diagrams will show:

| First Layer from the Top. | | | | | Second Layer from the Top. | | | | | Third Layer from the Top. | | | | | Fourth Layer from the Top. | | | |
|---|---|---|---|---|---|---|---|---|---|---|---|---|---|---|---|---|---|---|
| 1 | 48 | 32 | 49 | | 63 | 18 | 34 | 15 | | 62 | 19 | 35 | 14 | | 4 | 45 | 29 | 52 |
| 60 | 21 | 37 | 12 | | 6 | 43 | 27 | 54 | | 7 | 42 | 26 | 55 | | 57 | 24 | 40 | 9 |
| 56 | 25 | 41 | 8 | | 10 | 39 | 23 | 58 | | 11 | 38 | 22 | 59 | | 53 | 28 | 44 | 5 |
| 13 | 36 | 20 | 61 | | 51 | 30 | 46 | 3 | | 50 | 31 | 47 | 2 | | 16 | 33 | 17 | 64 |

The same sum 130 here comes out not less than 52 times; viz. in the first place from the 16 rows from left to right, secondly from the 16 rows from the front to the back, thirdly from the 16 rows counting from the top to the bottom, and lastly from the 4 rows which join each two opposite corners of the cube, namely from the rows: 1, 43, 22, 64; 49, 27, 38, 16; 13, 39, 26, 52; 61, 23, 42, 4.

For a cube with 5 compartments in each edge the arrangement of the figures can so be made that all the 75 rows parallel to any and every edge, all the 30 rows lying in any diagonal of a square, and all the 4 rows forming any principal diagonal shall have one and the same summation, 315.

Just as the magic squares of an odd number of cells could be formed with the aid of *two* auxiliary squares, so also odd-numbered magic cubes can be constructed with the help of *three* auxiliary cubes.

| First Layer from Top. | | | | | Second Layer from Top. | | | | | Third Layer from Top. | | | | |
|---|---|---|---|---|---|---|---|---|---|---|---|---|---|---|
| 121 | 27 | 83 | 14 | 70 | 2 | 58 | 114 | 45 | 96 | 33 | 89 | 20 | 71 | 102 |
| 10 | 61 | 117 | 48 | 79 | 36 | 92 | 23 | 54 | 110 | 67 | 123 | 29 | 85 | 11 |
| 44 | 100 | 1 | 57 | 113 | 75 | 101 | 32 | 88 | 19 | 76 | 7 | 63 | 119 | 50 |
| 53 | 109 | 40 | 91 | 22 | 84 | 15 | 66 | 122 | 28 | 115 | 41 | 97 | 3 | 59 |
| 87 | 18 | 74 | 105 | 31 | 118 | 49 | 80 | 6 | 62 | 24 | 55 | 106 | 37 | 93 |

| Fourth Layer from Top. | | | | | Lowest Layer. | | | | |
|---|---|---|---|---|---|---|---|---|---|
| 64 | 120 | 46 | 77 | 8 | 95 | 21 | 52 | 108 | 39 |
| 98 | 4 | 60 | 111 | 42 | 104 | 35 | 86 | 17 | 73 |
| 107 | 38 | 94 | 25 | 51 | 13 | 69 | 125 | 26 | 82 |
| 16 | 72 | 103 | 34 | 90 | 47 | 78 | 9 | 65 | 116 |
| 30 | 81 | 12 | 68 | 124 | 56 | 112 | 43 | 99 | 5 |

In this manner the preceding magic cube of 5 times 5 times 5 compartments is formed, in which, it may be additionally noticed, the middle number between 1 and 125, namely 63, is placed in the central compartment; by which arrangement the attainment of the sum of 315 is assured in the four principal diagonals and the 30 sub-diagonals. The condition attained in the magic squares, that the diagonal-pairs parallel to the sub-diagonals also shall give the sum 315 is not attainable in this case but is so in the case of higher numbers of compartments.

## CONCLUSION.

Musing on such problems as are the magic squares is fascinating to thinkers of a mathematical turn of mind. We take delight in discovering a harmony that abides as an intrinsic quality in the forms of our thought. The problems of the magic squares are playful puzzles, invented as it seems for mere pastime and sport. But there is a deeper problem underlying all these little riddles, and this deeper problem is of a sweeping significance. It is the philosophical problem of the world-order.

The formal sciences are creations of the mind. We build the sciences of mathematics, geometry, and algebra with our conception of pure forms which are abstract ideas. And the same order that prevails in these mental constructions permeates the universe, so that an old philosopher, overwhelmed with the grandeur of law, imagined he heard its rhythm in a cosmic harmony of the spheres.

# THE FOURTH DIMENSION.

## MATHEMATICAL AND SPIRITUALISTIC.

I.

### INTRODUCTORY.

THE tendency to generalise long ago led mathematicians to extend the notion of three-dimensional space, which is the space of sensible representation, and to define aggregates of points, or spaces, of more than three dimensions, with the view of employing these definitions as useful means of investigation. They had no idea of requiring people to imagine four-dimensional things and worlds, and they were even still less remote from requiring them to believe in the real existence of a four-dimensioned space. In the hands of mathematicians this extension of the notion of space was a mere means devised for the discovery and expression, by shorter and more convenient ways, of truths applicable to common geometry and to algebra operating with more than three unknown quantities. At this stage, however, the spiritualists came in, and coolly took possession of this private property of the mathematicians. They were in great perplexity as to where they should put the spirits of the dead. To give them a place in the world accessible to our senses was not exactly practicable. They were compelled, therefore, to look around after some *terra incognita*, which should oppose to the spirit of research inborn in humanity an insuperable barrier. The abiding-place of the spirits had perforce to be inaccessible to the senses and full of mystery to the mind. This property the four-dimensioned space of the mathematicians possessed. With an intellectual perversity which science has no idea of, these spiritualists boldly asserted, first, that the whole world was situated in a four-dimensioned space as a plane might be situated in the space familiar to us, secondly, that the spirits of the dead lived in such a four-dimensioned space, thirdly, that these spirits could accordingly act upon the world and, consequently, upon the human beings resident in it, exactly as we three-dimensioned creatures can produce effects upon things that are two-dimensioned; for example, such effects as that produced when we shatter a lamina of ice, and so influence some possibly existing two-dimensioned *ice*-world.

Since spiritualism, under the leadership of a Leipsic Professor, Zöllner, thus proclaimed the existence of a four-dimensioned space, this notion, which the mathematicians are thoroughly master of,—for in all their operations with it, though they have forsaken the path of actual representability, they have never left that of the truth,—this notion has also passed into the heads of lay persons who have used it as a catchword, ordinarily without having any clear idea of what they or any one else mean by it. To clear up such ideas and to correct the wrong impressions of cultured people who have not a technical mathematical training, is the purpose of the following pages. A similar elucidation was aimed at in the tracts which Schlegel (Riemann, Berlin, 1888) and Cranz (Virchow-Holtzendorff's Sammlung, Nos. 112 and 113) have published on the so-called fourth dimension. Both treatises possess indubitable merits, but their methods of presentation are in many respects too concise to give to lay minds a profound comprehension of the subject. The author, accordingly, has been able to add to the reflections which these excellent treatises offer, a great deal that appears to him necessary for a thorough explanation in the minds of non-mathematicians of the notion of the fourth dimension.

## II.

### THE CONCEPT OF DIMENSION.

Many text-books of stereometry begin with the words: "Every body has three dimensions, length, breadth, and thickness." If we should ask the author of a book of this description to tell us the length, breadth, and thickness of an apple, of a sponge, or of a cloud of tobacco smoke, he would be somewhat perplexed and would probably say, that the definition in question referred to something different. A cubical box, or some similar structure, whose angles are all right angles and whose bounding surfaces are consequently all rectangles is the only body of which it can at all be unmistakably asserted that there are three principal directions distinguishable in it, of which any one can be called the length, any other the breadth, and any third the thickness. We thus see that the notions of length, breadth, and thickness are not sufficiently clear and universal to enable us to derive from them any idea of what is meant when it is said that every body possesses three dimensions, or that the space of the world is three-dimensional.

This distinction may be made sharper and more evident by the following considerations : We have, let us suppose, a straight line on which a point is situated, and the problem is proposed to determine the position of the point on the line in an unequivocal manner. The simplest way to solve this is, to state how far the point is removed in the one or the other direction from some given fixed point; just as in a thermometer the position of the surface of the mercury is given by a statement of its distance in the direction of cold or heat from a predetermined fixed point—the point of freezing water. To state, therefore, the position of a point on a straight line, the sole datum necessary is a single number, if beforehand we have fixed upon some standard line, like the centimetre, and some definite point to which we give the value zero, and have also previously decided in what direction from the zero-point, points must be situated whose position is expressed by positive numbers, and also in what direction those must lie whose position is expressed by negative numbers. This last-mentioned fact, that a *single* number is sufficient to determine the place of a point in a straight line, is the real reason why we attribute to the straight line or to any part of it a single dimension.

More generally, we call every totality or system, of infinitely numerous things, *one*-dimensional, in which *one* number is all that is requisite to determine and mark out any particular one of these things from among the entire totality. Thus, time is one-dimensional. We, as inhabitants of the earth, have naturally chosen as our unit of time, the period of the rotation of the earth about its axis, namely, the day, or a definite portion of a day. The zero-point of time is regarded in Christian countries as the year of the birth of Christ, and the positive direction of time is the time *subsequent* to the birth of Christ. These data fixed, all that is necessary to establish and distinguish any definite point of time amid the infinite totality of all the points of time, *is a single number.* Of course this number need not be a whole number, but may be made up of the sum of a whole number and a fraction in whose numerator and denominator we may have numbers as great as we please. We may, therefore, also say that the totality of all conceivable numerical magnitudes, or of only such as are greater than one definite number and smaller than some other definite number, is one-dimensional.

We shall add here a few additional examples of one-dimensional magnitudes presented by geometry. First, the circumference of a circle is a one-dimensional magnitude, as is every curved line, whether it returns into itself or not. Further, the totality of all equilateral triangles which stand on the same base is one-dimensional, or the totality of all circles that can be described through two fixed points. Also, the totality of all conceivable cubes will be seen to be one-dimensional, provided they are distinguished, not with respect to position, but with respect to magnitude.

In conformity with the fundamental ideas by which we define the notion of a one-dimensional manifoldness, it will be seen that the attribute *two*-dimensional must be applied to all totalities of things in which *two* numbers are necessary (and sufficient) to distinguish any determinate individual thing amid the totality. The simplest two-dimensioned complex which we know of is the plane. To determine accurately the position of a point in a plane, the simplest way is to take two axes at right angles to each other, that is, fixed straight lines, and then to specify the distances by which the point in question is removed from each of these axes.

This method of determining the position of a point in a plane suggested to the celebrated philosopher and mathematician Descartes the fundamental idea of analytical geometry, a branch of mathematics in which by the simple artifice of ascribing to every point in a plane two numerical values, determined by its distances from the two axes above referred to, planimetrical considerations are transformed into algebraical. So, too, all kinds of curves that graphically represent the dependence of things on time, make use of the fact that the totality of the points in a plane is two-dimensional. For example, to represent in a graphical form the increase in the population of a city, we take a horizontal axis to represent the time, and a perpendicular one to represent the numbers which are the measures of the population. Any two lines, then, whose lengths practical considerations determine, are taken as the unit of time, which we may say is a year, and as the unit of population, which we will say is one thousand. Some definite year, say 1850, is fixed upon as the zero point. Then, from all the equally distant points on the horizontal axis, which points stand for the years, we proceed in directions parallel to the other axis, that is, in the perpendicular direction, just so much upwards as the numbers which stand for the population of that year require. The terminal points so reached, or the curve which runs through these terminal points, will then present a graphic picture of the rates of increase of the population of the town in the different years. The rectangular axes of Descartes are employed in a similar way for the construction of barometer curves, which specify for the different localities of a country the amount of variation of the atmospheric pressure during any period of time. Immediately next to the plane the surface of the earth will be recognised as a two-dimensional aggregate of points. In this case geographical latitude and longitude supply the two numbers that are requisite accurately to determine the position of a point. Also, the totality of all the possible straight lines that can be drawn through any point in space is two-dimensional, as we shall best understand if we picture to ourselves a plane which is cut in a point by each of these straight lines and then remember that by such a construction every point on the plane will belong to some one line and, *vice versa*, a line to every point, whence it follows that the totality of all the straight lines, or, as we may call them, rays, which pass through the point assigned are of the same dimensions as the totality of the points of the imagined plane.

The question might be asked, In what way and to what extent in this case is the specification of *two* numbers requisite and sufficient to determine amid all the rays which pass through the specified point a definite individual ray? To get a clear idea of the problem here involved, let us imagine the ray produced far into the heavens where some quite definite point will correspond to it. Now, the position of a point in the heavens depends, as does the position of a point on all spherical surfaces, on two numbers. In the heavens these two numbers are ordinarily supplied by the two angles called altitude, or the distance above the plane of the horizon, and azimuth, or the angular distance between the circle on which the altitude is measured and the meridian of the observer. It will be seen thus that the totality of all the luminous rays that an eye, conceived as a point, can receive from the outer world is two-dimensional, and also that a luminous point emits a two-dimensional group of luminous rays. It will also be observed, in connexion with this example, that the two-dimensional totality of all the rays that can be drawn through a point in space is something different from the totality of the rays that pass through a point but are required to lie in a given plane. Such a group of objects as the last-named one is a one-dimensional totality.

Now that we have sufficiently discussed the attributes that are characteristic of one and two-dimensional aggregates, we may, without any further investigation of the subject, propose the following definition, that, generally, *an n-dimensional totality of infinitely numerous things is such that the specification of n numbers is necessary and sufficient to indicate definitely any individual amid all the infinitely numerous individuals of that totality.*

Accordingly, the point-aggregate made up of the world-space which we inhabit, is a three-dimensional totality. To get our bearings in this space and to define any determinate point in it, we have simply to lay through any point which we take as our zero-point three axes at right angles to each other, one running from right to left, one backwards and forwards, and one upwards and downwards. We then join each two of these axes by a plane and are enabled thus to specify the position of every point in space by the three perpendicular distances by which the point in question is removed in a

positive or negative sense from these three planes.  It is customary
to denote the numbers which are the measures of these three distances by $x$, $y$, and $z$, the positive $x$, positive $y$, and positive $z$ ordinarily being reckoned in the right hand, the hitherward, and the
upward directions from the origin.  If now, with direct reference to
this fundamental axial system, any particular specification of $x$, $y$,
and $z$ be made, there will, by such an operation, be cut out and isolated from the three-dimensional manifoldness of all the points of
space a totality of less dimensions.  If, for example, $z$ is equal to
seven units or measures, this is equivalent to a statement that only
the two-dimensional totality of the points is meant, which constitute the plane that can be laid at right angles to the upward-passing
$z$-axis at a distance of seven measures from the zero-point.  Consequently, every imaginable equation between $x$, $y$, and $z$ isolates and
defines a two-dimensional aggregate of points.  If two different equations obtain between $x$, $y$, and $z$, two such two-dimensional totalities
will be isolated from among all the points of space.  But as these
last must have some one-dimensional totality in common, we may
say that the co-existence of two equations between $x$, $y$, and $z$ defines
a one-dimensional totality of points, that is to say a straight line, a
line curved in a plane, or even, perhaps, one curved in space.  It
is evident from this that the introduction of the three axes of reference forms a bridge between the theory of space and the theory of
equations involving three variable quantities, $x$, $y$, $z$.  The reason
that the theory of space cannot thus be brought into connection
with algebra in general, that is, with the theory of indefinitely numerous equations, but only with the algebra of three quantities, $x$,
$y$, $z$, is simply to be sought in the fact that space, as we picture it,
can have only three dimensions.

We have now only to supply a few additional examples of $n$-
dimensional totalities.  All particles of air are four-dimensional in
magnitude when in addition to their position in space we also consider the variable densities which they assume, as they are expressed
by the different heights of the barometer in the different parts of the
atmosphere.  Similarly, all conceivable spheres in space are four-
dimensional magnitudes, for their centres form a three-dimensional

point-aggregate, and around each centre there may be additionally conceived a one-dimensional totality of spheres, the radii of which can be expressed by every numerical magnitude from zero to infinity. Further, if we imagine a measuring-stick of invariable length to assume every conceivable position in space, the positions so obtained will constitute a five-dimensional aggregate. For, in the first place, one of the extremities of the measuring stick may be conceived to assume a position at every point of space, and this determines for one extremity alone of the stick a three-dimensional totality of positions; and secondly, as we have seen above, there proceeds from every such position of this extremity a two-dimensional totality of directions, and by conceiving the measuring-stick to be placed lengthwise in every one of these directions we shall obtain all the conceivable positions which the second extremity can assume, and consequently, the dimensions must be 3 plus 2 or 5. Finally, to find out how many dimensions the totality of all the possible positions of a square, invariable in magnitude, possesses, we first give one of its corners all conceivable positions in space, and we thus obtain three dimensions. One definite point in space now being fixed for the position of one corner of the square, we imagine drawn through this point all possible lines, and on each we lay off the length of the side of the square and thus obtain two additional dimensions. Through the point obtained for the position of the second corner of the square we must now conceive all the possible directions drawn that are perpendicular to the line thus fixed, and we must lay off once more on each of these directions the side of the square. By this last determination the dimensions are only increased by one, for only one one-dimensional totality of perpendicular directions is possible to one straight line in one of its points. Three corners of the square are now fixed and therewith the position of the fourth also is uniquely determined. Accordingly, the totality of all equal squares which differ from one another only by their position in space, constitutes a manifoldness of six dimensions.

### III.

### THE INTRODUCTION OF THE NOTION OF FOUR-DIMENSIONAL POINT-AGGREGATES, PERMISSIBLE.

In the preceding section it was shown that we can conceive not only of manifoldnesses of one, two, and three dimensions, but also of manifoldnesses of *any* number of dimensions. But it was at the same time indicated that our world-space, that is, the totality of all conceivable *points* that differ only in respect of position, cannot in agreement with our notions of things possess more than three dimensions. But the question now arises, whether, if the progress of science tends in such a direction, it is permissible to extend the notion of space by the introduction of point-aggregates of more than three dimensions, and to engage in the study of the properties of such creations, although we know that notwithstanding the fact that we may conceptually establish and explore such aggregates of points, yet we cannot picture to ourselves these creations as we do the spatial magnitudes which surround us, that is, the regular three-dimensional aggregates of points.

To show the reader clearly that this question must be answered in the affirmative, that the extension of our notion of space is permissible, although it leads to things which we cannot perceive by our senses, I may call the reader's attention to the fact that in arithmetic we are accustomed from our youth upwards to extensions of ideas, which, accurately viewed, as little admit of graphic conception as a four-dimensional space, that is, a point-aggregate of four dimensions. By his senses man first reaches only the idea of whole numbers—the results of counting. The observation of primitive peoples* and of children clearly proves that the essential decisive factors of counting are these three : First, we abstract, in the counting of things, completely from the individual and characteristic attributes of these things, that is, we consider them as homogeneous. Second, we associate individually with the things which we count

---

* See the essay *Notion and Definition of Number* in this collection.

other homogeneous things. These other things are even now, among uncivilised peoples, the ten fingers of the two hands. They may, however, be simple strokes, or, as in the case of dice and dominoes, black points on a white background. Third, we substitute for the result of this association some concise symbol or word ; for example, the Romans substituted for three things counted, three strokes placed side by side, namely : III ; but for greater numbers of things they employed abbreviated signs. The Aztecs, the original inhabitants of Mexico, had time enough, it seems, to express all the numbers up to nineteen by equal circles placed side by side. They had abbreviated signs only for the numbers 20, 400, 8000, and so forth. In speaking, some one same sound might be associated with the things counted ; but this method of counting is nowadays employed only by clocks: the languages of men since prehistoric times have fashioned concise words for the results of the association in question. From the notion of number, thus fixed as the result of counting, man reached the notion of the addition of two numbers, and thence the notion that is the inverse of the last process, the notion of subtraction. But at this point it clearly appears that not every problem which may be propounded is soluble ; for there is no number which can express the result of the subtraction of a number from one which is equally large or from one which is smaller than itself. The primary school pupil who says that 8 from 5 "won't go" is perfectly right from his point of view. For there really does not exist any result of counting which added to eight will give five.

If humanity had abided by this point of view and had rested content with the opinion that the problem "5 minus 8" is not solvable, the science of arithmetic would never have received its full development, and humanity would not have advanced as far in civilisation as it has. Fortunately, men said to themselves at this crisis: "If 5 minus 8 won't go, we'll *make it go;* if 5 minus 8 does not possess an intelligible meaning, we will simply give it one." As a fact, things which have not a meaning always afford men a pleasing opportunity of investing them with one. The question is, then, what significance is the problem "5 minus 8" to be invested with?

The most natural and therefore the most advantageous solution undoubtedly is to abide by the original notion of subtraction as
the inverse of addition, and to make the significance of 5 minus 8
such, that for 5 minus 8 plus 8 we shall get our original minuend 5.
By such a method all the rules of computation which apply to real
differences will also hold good for unreal differences, such as 5 minus
8. But it then clearly appears that all forms expressive of differences in which the numbers that stand before the minus symbol
are less by an equal amount than those which follow it may be regarded as equal; so that the simplest course seems to be to introduce as the common characteristic of all equal differential forms of
this description a common sign, which will indicate at the same
time the difference of the two numbers thus associated. Thus it
came about that for 5 minus 8, as well as for every differential form
which can be regarded as equal thereto, the sign "— 3" was introduced. But in calling differential forms of this description numbers, the notion of number was extended and a new domain was
opened up, namely the domain of negative numbers.

In the further development of the science of arithmetic, through
the operation of division viewed as the inverse of multiplication, a
second extension of the idea of number was reached, namely, the
notion of fractional numbers as the outcome of divisions that had
led to numbers hitherto undefined. We find, thus, that the science
of arithmetic throughout its whole development has strictly adhered
to the principle of conformity and consistency and has invested
every association of two numbers, which before had no significance,
by the introduction of new numbers, with a real significance, such
that similar operations in conformity with exactly the same rules
could be performed with the new numbers, viewed as the results of
this association, as with the numbers which were before known and
perfectly defined. Thus the science proceeded further on its way and
reached the notions of irrational, imaginary, and complex numbers.

The point in all this, which the reader must carefully note, is,
that all the numbers of arithmetic, with the exception of the positive whole numbers, are artificial products of human thought, invented to make the language of arithmetic more flexible, and to

accelerate the progress of science. All these numbers lack the attributes of representability.

No man in the world can picture to himself "minus three trees." It is possible, of course, to know that when three trees of a garden have been cut down and carried away, three are missing, and by substituting for "missing" the inverse notion of "added," we may say, perhaps, that "minus three trees" are added. But this is quite different from the feat of imagining a negative number of trees. We can only picture to ourselves a number of trees that results from actual counting, that is, a positive whole number. Yet, notwithstanding all this, people had not the slightest hesitation in extending the notion of number. Exactly so must it be permitted us in geometry to extend the notion of space, even though such an extension can only be mentally defined and can never be brought within the range of human powers of representation.

In mathematics, in fact, the extension of any notion is admissible, provided such extension does not lead to contradictions with itself or with results which are well established. Whether such extensions are necessary, justifiable, or important for the advancement of science is a different question. It must be admitted, therefore, that the mathematician is justified in the extension of the notion of space as a point-aggregate of three dimensions, and in the introduction of space or point-aggregates of more than three dimensions, and in the employment of them as means of research. Other sciences also operate with things which they do not know exist, and which, though they are sufficiently defined, cannot be perceived by our senses. For example, the physicist employs the ether as a means of investigation, though he can have no sensory knowledge of it. The ether is nothing more than a means which enables us to comprehend mechanically the effects known as action at a distance and to bring them within the range of a common point of view. Without the assumption of a material which penetrates everything, and by means of whose undulations impulses are transmitted to the remotest parts of space, the phenomena of light, of heat, of gravitation, and of electricity would be a jumble of isolated and unconnected mysteries. The assumption of an ether, however, comprises in a systematic scheme all these isolated events, facilitates our mental control of the phenomena of nature, and enables us to produce these phenomena at will.    But it must not be forgotten in such reflexions that the ether itself is even a greater problem for man, and that the ether-hypothesis does not solve the difficulties of phenomena, but only puts them in a unitary conceptual shape.    Notwithstanding all this, physicists have never had the least hesitation in employing the ether as a means of investigation.    And as little do reasons exist why the mathematicians should hesitate to investigate the properties of a four-dimensioned point-aggregate, with the view of acquiring thus a convenient means of research.

IV.

### THE INTRODUCTION OF THE IDEA OF FOUR-DIMENSIONED POINT-AGGREGATES OF SERVICE TO RESEARCH.

From the concession that the mathematician has the right to define and investigate the properties of point-aggregates of more than three dimensions, it does not necessarily follow that the introduction of an idea of this description is of value to science.    Thus, for example, in arithmetic, the introduction of operations which spring from involution, as involution and its two inverse operations proceed from multiplication, is undoubtedly permitted.    Just as for "$a$ times $a$ times $a$" we write the abbreviated symbol "$a^3$," (which we read, $a$ to the third power,) and investigate in detail the operation of involution thus defined, so we might also introduce some shorthand symbol for "$a$ to the $a^{th}$ power to the $a^{th}$ power" and thus reach an operation of the fourth degree, which would regard $a$ as a passive number and the number 3, or any higher number, as the active number, that is, as the number which indicates how often $a$ is taken as the base of a power whose exponent may be $a$, or "$a$ to the $a^{th}$," or "$a$ to the $a^{th}$ to the $a^{th}$ power."

But the introduction of such an operation of the fourth degree has proved itself to be of no especial value to mathematics.    And the reason is that in the operation of involution the law of commutation does not hold good.    In addition, the numbers to be added may be interchanged and the introduction of multiplication is therefore of great value.  So, also, in multiplication the numbers which are combined, that is, the factors, may be changed about in any way, and thus the introduction of involution is of value.  But in involution the base and the exponent cannot be interchanged, and consequently the introduction of any higher operation is almost valueless.

But with the introduction of the idea of point-aggregates of multiple dimensions the case is wholly different.   The innovation in question has proved itself to be not only of great importance to research, but the progress of science has irresistibly forced investigators to the introduction of this idea, as we shall now set forth in detail.

In the first place, algebra, especially the algebraical theory of systems of eqations, derives much advantage from the notion of multi-dimensioned spaces.  If we have only three unknown quantities, $x$, $y$, $z$, the algebraical questions which arise from the possible problems of this class admit, as we have above seen, of geometrical representation to the eye.   Owing to this possibility of geometrical representation, some certain simple geometrical ideas like "moving," "lying in," "intersecting," and so forth, may be translated into algebraical events.   Now, no reason exists why algebra should stop at three variable quantities ; it must in fact take into consideration any number of variable quantities.

For purposes of brevity and greater evidentness, therefore, it is quite natural to employ geometrical forms of speech in the consideration of more than three variables.  But when we do this, we assume, perhaps without really intending to do so, the idea of a space of more than three dimensions. If we have four variable quantities, $x$, $y$, $z$, $u$, we arrive, by conceiving attributed to each of these four quantities every possible numerical magnitude, at a four-dimensioned manifoldness of numerical quantities, which we may just as well regard as a four-dimensioned aggregate of points.   Two equations which exist on this supposition between $x$, $y$, $z$, and $u$, define two three-dimensioned aggregates of points, which intersect, as we may briefly say, in a two-dimensioned aggregate of points, that is, in a surface ; and so on.   In a somewhat different manner the determination of the contents of a square or a cube by the involution of a number which stands for the length of its sides, leads to the notion of four-dimensioned structures, and, consequently, to the notion of a four-dimensioned point-space. When we note that $a^2$ stands for the contents of a square, and $a^3$ for the content of a cube, we naturally inquire after the contents of a structure which is produced from the cube as the cube is produced from the square and which also will have the contents $a^4$. We cannot, it is true, clearly picture to ourselves a structure of this description, but we can, nevertheless, establish its properties with mathematical exactness.* It is bounded by 8 cubes just as the cube is bounded by 6 squares; it has 16 corners, 24 squares, and 32 edges, so that from every corner 4 edges, 6 squares, and 4 cubes proceed, and from every edge 3 squares and 3 cubes.

Yet despite the great service to algebra of this idea of multi-dimensioned space, it must be conceded that the conception although convenient is yet not indispensable. It is true, algebra is in need of the idea of multiple dimensions, but it is not so absolutely in need of the idea of *point* aggregates of multiple dimensions.

This notion is, however, necessary and serviceable for a profound comprehension of geometry. The system of geometrical knowledge which Euclid of Alexandria created about three hundred years before Christ, supplied during a period of more than two thousand years a brilliant example of a body of conclusions and truths which were mutually consistent and logical. Up to the present century the idea of elementary geometry was indissolubly bound up with the name of Euclid, so that in England where people adhered longest to the rigid deductive system of the Grecian mathematician, the task of "learning geometry" and "reading Euclid" were until a few years ago identical. Every proposition of this Euclidean system rests on other propositions, as one building-stone in a house rests upon another. Only the very lowest stones, the foundations, were without supports. These are the axioms or fun-

---

* Victor Schlegel, indeed, has made models of the three-dimensional nets of all the six structures which correspond in four-dimensioned space to the five regular bodies of our space, in an analogous manner to that by which we draw in a plane the nets of five regular bodies. Schlegel's models are made by Brill of Darmstadt.

damental propositions, truths on which all other truths are, directly or indirectly, founded, but which themselves are assumed without demonstration as self-evident.

But the spirit of mathematical research grew in time more and more critical, and finally asked, whether these axioms might not possibly admit of demonstration. Especially was a rigid proof sought for the eleventh * axiom of Euclid, which treats of parallels.

After centuries of fruitless attempts to prove Euclid's eleventh axiom, Gauss, and with him Bolyai and Lobachévski, Riemann, and Helmholtz, finally stated the decisive reasons why any attempt to prove the axiom of the parallels must necessarily be futile. These reasons consist of the fact that though this axiom holds good enough in the world-space such as we do, and can conceive it, yet three-dimensioned spaces are ideally conceivable though not capable of mental representation, where the axiom does not hold good. The axiom was thus shown to be a mere fact of *observation*, and from that time on there could no longer be any thought of a deductive demonstration of it. In view of the intimate connection, which both in an historical and epistemological point of view exists between the extension of the concept of space and the critical examination of the axioms of Euclid, we must enter at somewhat greater length into the discussion of the last mentioned propositions.

Of the axioms which Euclid lays at the foundation of his geometry, only the following three are really geometrical axioms :

*Eighth axiom:* Magnitudes which coincide with one another are equal to one another.

*Eleventh axiom:* If a straight line meet two straight lines so as to make the two interior angles on the same side of it taken together less than two right angles, these straight lines, being continually produced, shall at length meet on that side on which are the angles which are less than two right angles.

*Twelfth axiom:* Two straight lines cannot inclose a [finite] space.

The numerous proofs which in the course of time were adduced

---

* Also called the twelfth axiom, also the fifth postulate.— *Tr.*

in demonstration of these axioms, especially of the eleventh, all turn out on close examination to be pseudo-proofs. Legendre drew attention to the fact that either of the following axioms might be substituted for the eleventh:

*a*) Given a straight line, there can be drawn through a point in the same plane with that line, one and one line only which shall not intersect the first (parallels) however far the two lines may be produced;

*b*) If two parallel lines are cut by a third straight line, the interior alternate angles will be equal.

*c*) The sum of the angles of a triangle is equal to two right angles, that is, to the angle of a straight line or to 180°.

By the aid of any one of these three assertions, the eleventh axiom of Euclid may be proved, and, *vice versa*, by the aid of the latter each of the three assertions may be proved, of course with the help of the other two axioms, eight and twelve. The perception that the eleventh axiom does not admit of demonstration without the employment of one of the foregoing substitutes may best be gained from the consideration of congruent figures. Every reader will remember from his first instruction in geometry that the congruence of two triangles is demonstrated by the superposition of one triangle on the other and by then ascertaining whether the two completely coincide, no assumptions being made in the determination except those above mentioned.

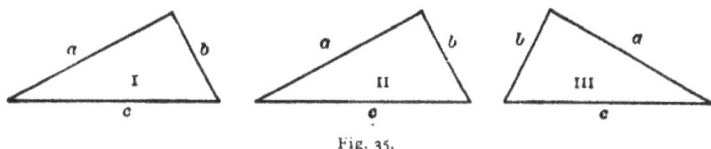

Fig. 35.

In the case of triangles which are congruent, as are I and II in the preceding cut, this coincidence may be effected by the simple *displacement* of one of the triangles; so that even a two-dimensional being, supposed to be endowed with powers of reasoning, but only capable of picturing to itself motions within a plane, also might convince itself that the two triangles I and II could be made to coincide. But a being of this description could not convince itself in like manner of the congruence of triangles I and III. It would discover the equality of the three sides and the three angles, but it could never succeed in so superposing the two triangles on each other as to make them coincide. A three-dimensional being, however, can do this very easily. It has simply to turn triangle I about one of its sides and to shove the triangle, thus brought into the position of its reflexion in a mirror, into the position of triangle III. Similarly, triangles II and III may be made to coincide by moving either out of the plane of the paper around one of its sides as axis and turning it until it again falls in the plane of the paper. The triangle thus turned over can then be brought into the position of the other.

Later on we shall revert to these two kinds of congruence: "congruence by displacement" and "congruence by circumversion." For the present we will start from the fact that it is always possible within the limits of a plane to take a triangle out of one position and bring it into another without altering its sides and angles. The question is, whether this is only possible in the plane, or whether it can also be done on other surfaces.

We find that there are certain surfaces in which this is possible, and certain others in which it is not. For instance, it is impossible to move the triangle drawn on the surface of an egg into some other position on the egg's surface without a distension or contraction of some of the triangle's parts. On the other hand, it is quite possible to move the triangle drawn on the surface of a sphere into any other position on the sphere's surface without a distension or contraction of its parts. The mathematical reason of this fact is, that the surface of a sphere, like the plane, has everywhere the same curvature, but that the surface of an egg at differ-ent places has different curvatures. Of a plane we say that it has everywhere the curvature zero; of the surface of a sphere we say it has everywhere a positive curvature, which is greater in proportion as the radius is smaller. There are surfaces also which have a constant negative curvature; these surfaces exhibit at every point in directions proceeding from the same side a partly concave and a partly convex structure, somewhat like the centre of a saddle.

There is no necessity of our entering in any detail into the character and structure of the last-mentioned surfaces.

Intimately related with the plane, however, are all those surfaces, which, like the plane, have the curvature zero; in this category belong especially cylindrical surfaces and conical surfaces. A sheet of paper of the form of the sector of a circle may, for example, be readily bent into the shape of a conical surface. If two congruent triangles, now, be drawn on the sheet of paper, which may by displacement be translated the one into the other, these triangles will, it is plain, also remain congruent on the conical surface; that is, on the conical surface also we may displace the one into the other; for though a bending of the figures will take place, there will be no distension or contraction. Similarly, there are surfaces which, like the sphere, have everywhere a constant positive curvature. On such surfaces also every figure can be transferred into some other position without distension or contraction of its parts. Accordingly, on all surfaces thus related to the plane or sphere, the assumption which underlies the eighth axiom of Euclid, that it is possible to transfer into any new position any figure drawn on such surfaces without distortion, holds good.

The eleventh axiom in its turn also holds good on all surfaces of constant curvature, whether the curvature be zero or positive; only in such instances instead of "straight" line we must say "shortest" line. On the surface of a sphere, namely, two shortest lines, that is, arcs of two great circles, always intersect, no matter whether they are produced in the direction of the side at which the third arc of a great circle makes with them angles less than two right angles, or, in the direction of the other side, where this arc makes with them angles of more than two right angles. On the plane, however, two straight lines intersect only on the side where a third straight line that meets them makes with them interior angles less than two right angles.

The twelfth axiom of Euclid, finally, only holds good on the plane and on the surfaces related to it, but not on the sphere or other surfaces which, like the sphere, have a constant positive curvature. This also accounts for the fact that one of the three postulates which we regarded as substitutes for the eleventh axiom, though valid for the plane, is not true for the surface of a sphere; namely, the postulate that defines the sum of the angles of a triangle. This sum in a plane triangle is two right angles; in a spherical triangle it is more than two right angles, the spherical triangle being greater, the greater the excess the sum of its angles is above two right angles. It will be seen, from these considerations, that in geometries in which curved surfaces and not fixed planes are studied, the axioms of Euclid are either all or partially false.

The axioms of geometry thus having been revealed as facts of experience, the question suggested itself whether in the same way in which it was shown that different two-dimensional geometries were possible, also different three-dimensional systems of geometry might not be developed; and consequently what the relations were in which these might stand to the geometry of the space given by our senses and representable to our mind. As a fact, a three-dimensional geometry can be developed, which like the geometry of the surface of an egg will exclude the axiom that a figure or body can be transferred from any one part of space to any other and yet remain congruent to itself. Of a three-dimensional space in which such a geometry can be developed we say, that it has no constant measure of curvature.

The space which is representable to us, and which we shall henceforth call the *space of experience*, possesses, as our experiences without exception confirm, the especial property that every bodily thing can be transferred from any one part of it to any other without suffering in the transference any distension or any contraction. The space of experience, therefore, has a constant measure of curvature. The question, however, whether this measure of curvature is zero or positive, that is, whether the space of experience possesses the properties which in two-dimensional structures a plane possesses, or whether it is the three dimensional analogue of the surface of a sphere is one which future experience alone can answer. If the space of experience has a constant positive measure of curvature which is different from zero, be the difference ever so slight, a point which should move forever onward in a straight line, or, more ac-

curately expressed, in a shortest line, would sometime, though perhaps after having traversed a distance which to us is inconceivable,
ultimately have to arrive from the opposite direction at the place
from which it set out, just as a point which moves forever onward
in the same direction on the surface of a sphere must ultimately arrive at its starting point, the distance it traverses being longer the
greater the radius of the sphere or the smaller its curvature.

It will seem, at first blush, almost incredible, that the space of
experience possibly could have this property.  But an example,
which is the historical analogue of this modern transformation of our
conceptions, will render the idea less marvellous.  Let us transport
ourselves to the age of Homer.  At that time people believed that
the earth was a great disc surrounded on all sides by oceans which
were conceived to be in all directions infinitely great.  Indeed, for
the primitive man, who has never journeyed far from the place of
his birth, this is the most natural conception.  But imagine now that
some scholar had come, and had informed the Homeric hero Ulysses
that if he would travel forever on the earth in the same direction he
would ultimately come back to the point from which he started;
surely Ulysses would have gazed with as much astonishment upon
this scholar as we now look upon the mathematician who tells us
that it is possible that a point which moves forever onward in space
in the same direction may ultimately arrive at the place from which
it started.  But despite the fact that Ulysses would have regarded
the assertion of the scholar as false because contradictory to his
familiar conceptions, that scholar, nevertheless, would have been
right; for the earth is not a plane but a spherical surface.  So also
the mathematician may be right who bases this more recent strange
view on the possible fact that the space of experience may have a
measure of curvature which is not exactly zero but slightly greater
than zero.  If this were really the case, the *volume* of the space of
experience, though very large, would, nevertheless, be finite; just
as the real spherical surface of the earth as contrasted with the
Homeric plane surface is finite, having so and so many square miles.
When the objection is here made that a finiteness of space is totally
at variance with our modes of thought and conceptions, two ideas,

"infinitely great" and "unlimited," are confounded. All that is at
variance with our practical conceptions is that space can anywhere
have a boundary; not that it may possibly be of tremendous but
finite magnitude.

It will now be asked if we cannot determine by actual observation whether the measure of curvature of experiential space is exactly zero or slightly different therefrom. The theorem of the sum
of the angles of a triangle and the conclusions which follow from
this theorem do indeed supply us with a means of ascertaining this
fact. And the results of observation have been, that *the measure of
curvature of space is in all probability exactly equal to zero or if it is
slightly different from zero it is so little so that the technical means of
observation at our command and especially our telescopes are not competent to determine the amount of the deviation.* More, we cannot with
any certainty say.

All these reflections, to which the criticism of the hypotheses
that underlie geometry long ago led investigators, compel us to in-
stitute a comparison between the space of experience and other
three-dimensional aggregates of points (spaces), which we cannot
mentally represent but can in thought and word accurately define
and investigate. As soon, however, as we are fully implicated in the
task of accurately investigating the properties of three-dimensional
aggregates of points, we find ourselves similarly forced to regard
such aggregates as the component elements of a manifoldness of
more than three dimensions. In this way the exact criticism of
even ordinary geometry leads us to the abstract assumption of a
space of more than three dimensions. And as the extension of every
idea gives a clearer and more translucent form to the idea as it originally stood, here too the idea of multi-dimensioned aggregates
of points and the investigation of their properties has thrown a new
light on the truths of ordinary geometry and placed its properties
in clearer relief. Among the numerous examples which show how
the notion of a space of multiple dimensions has been of great service to science in the investigation of three-dimensional space, we
shall give one a place here which is within the comprehension of
non-mathematicians.

Imagine in a plane two triangles whose angles are denoted by pairs of numbers—namely, by 1-2, 1-3, 1-4, and 2-5, 3-5, 4-5. (See Fig. 36.) Let the two triangles so lie that the three lines which join the angles 1-2 and 2-5, 1-3 and 3-5, and 1-4 and 4-5 intersect at a point, which we will call 1-5. If now we cause the sides of the triangles which are opposite to these angles to intersect, it will be found that the points of intersection so obtained possess the peculiar property of lying all in one and the same straight line. The point of intersection of the connection 1-3 and 1-4 with the connection 4-5 and 3-5 may appropriately be called 3-4. Similarly, the point of intersection 2-4 is produced by the meeting of 4-5, 2-5 and 1-2, 1-4; and the point of intersection 2-3, by the meeting of 1-3, 1-2 and 3-5, 2-5. The statement, that the three points of intersection 3-4, 2-4, 2-3, thus obtained, lie in one straight line, can be proved by the principles of plane geometry only with difficulty and great circumstantiality. But by resorting to the three-dimensional space of experience, in which the plane of the drawing lies, the proposition can be rendered almost self-evident.

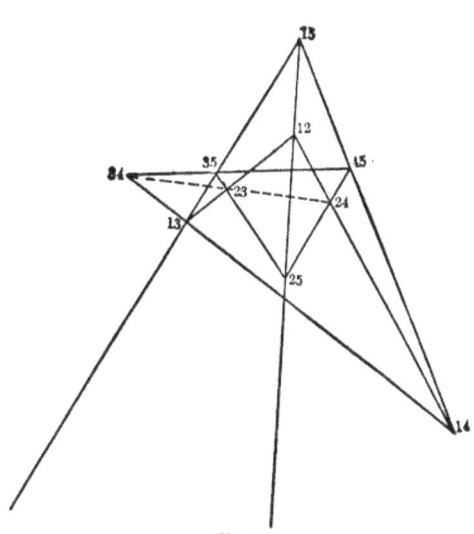

Fig. 36.

To begin with, imagine any five points in space which may be denoted by the numbers 1, 2, 3, 4, 5; then imagine all the possible ten straight lines of junction drawn between each two of these points, namely, 1-2, 1-3 . . . . 4-5; and finally, also, all the ten planes of junction of every three points described, namely, the plane 1-2-3, 1-2-4, . . . . 3-4-5. A spatial figure will thus be obtained, whose ten straight lines will meet some interposed plane in ten points whose relative positions are exactly those of the ten points above described.

Thus, for example, on this plane the points 1-2, 1-3, and 2-3 will lie in a straight line, for through the three spatial points 1, 2, 3, a plane can be drawn which will cut the plane of a drawing in a straight line. The reason, therefore, that the three points 3-4, 2-4, 2-3, also must ultimately lie in a straight line, consists in the simple fact that the plane of the three points 2, 3, 4, must cut the plane of the drawing in a straight line. The figure here considered consists of ten points of which sets of three so lie ten times in a straight line that conversely from every point also three straight lines proceed.

Now, just as this figure is a section of a complete three-dimensional pentagon, so another remarkable figure, of similar proper-

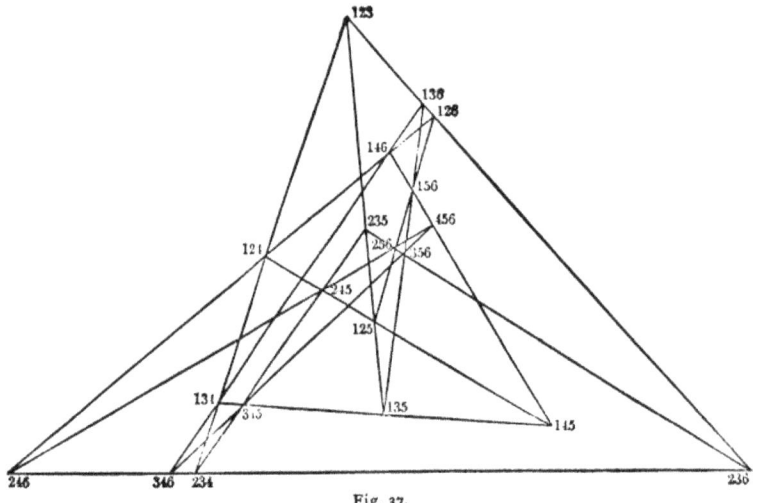

Fig. 37.

ties, may be obtained from the section of a figure of four-dimensioned space. Imagine six points, 1, 2, 3, 4, 5, 6, situated in this four-dimensioned space, and every three of them connected by a plane, and every four of them by a three-dimensioned space. We shall obtain thus twenty planes and fifteen three-dimensioned spaces which will cut the plane in which the figure is to be produced in twenty points and fifteen rays which so lie that each point sends out three rays and every ray contains four points. (See Fig. 37.) Figures of this description, which are so composed of points and rays that an equal number of rays proceed from every point and an equal

THE FOURTH DIMENSION.

number of points lie in every ray, are called *configurations*. Other configurations may, of course, be produced, by taking a different number of points and by assuming that the points taken lie in a space of different or even higher dimensions. The author of this article was the first to draw attention to configurations derived from spaces of higher dimensions. As we see, then, the notion of a space of more than three dimensions has performed an important service in the investigations of common plane geometry.

In conclusion, I should like to add a remark which Cranz makes regarding the application of the idea of multi-dimensioned space to theoretical chemistry. (See the treatise before cited.) In chemistry, the molecules of a compound body are said to consist of the atoms of the elements which are contained in the body, and these are supposed to be situated at certain distances from one another, and to be held in their relative positions by certain forces. At first, the centres of the atoms were conceived to lie in one and the same plane. But Wislicenus was led by researches in paralactic acid to explain the differences of isomeric molecules of the same structural formulæ by the different positions of the atoms in *space*. (Compare *La chimie dans l'espace* by van't Hoff, 1875, preface by J. Wislicenus). In fact four points can always be so arranged in space that every two of them may have any distance from each other; and the change of one of the six distances does not necessarily involve the alteration of any other.

But suppose our molecule consists of five atoms? Four of these may be so placed that the distance between any two of them can be made what we please. But it is no longer possible to give the fifth atom a position such that each of the four distances by which it is separated from the other atoms may be what we please. On the contrary, the fourth distance is dependent on the three remaining distances; for the space of experience has only three dimensions. If, therefore, I have a molecule which consists of five atoms I cannot alter the distance between two of them without at least altering some second distance. But if we imagine the centres of the atoms placed in a four-dimensioned space, this can be done; all the ten distances which may be conceived to exist between the five points will then be independent of one another. To reach the same result in the case of six atoms we must assume a five-dimensional space; and so on.

Now, if the independence of all the possible distances between the atoms of a molecule is absolutely required by theoretical chemical research, the science is really compelled, if it deals with molecules of more than four atoms, to make use of the idea of a space of more than three dimensions. This idea is, in this case, simply an instrument of research, just as are, also, the ideas of molecules and atoms—means designed to embrace in a perspicuous and systematic form the phenomena of chemistry and to discover the conditions under which new phenomena can be evoked. Whether a four-dimensioned space really exists is a question whose insolubility cannot prevent research from making use of the idea, exactly as chemistry has not been prevented from making use of the notion of atom, although no one really knows whether the things we call atoms exist or not.

v.

REFUTATION OF THE ARGUMENTS ADDUCED TO PROVE THE
EXISTENCE OF A FOUR-DIMENSIONED SPACE INCLUSIVE
OF THE VISIBLE WORLD.

The considerations of the preceding section will have convinced the cultured non-mathematician of the service which the theory of multi-dimensioned spaces has done, and bids fair to do, for geometrical research. In addition thereto is the consideration that every extension of one branch of mathematical science is a constant source of beneficial and helpful influence to the other branches. The knowledge, however, that mathematicians can employ the notion of four-dimensioned space with good results in their researches, would never have been sufficient to procure it its present popularity; for every man of intelligence has now heard of it, and, in jest or in earnest, often speaks of it. The knowledge of a four-dimensioned space did not reach the ears of cultured non-mathematicians until the consequences which the spiritualists fancied it was permissible to draw from this mathematical notion were publicly known. But it is a tremendous step from the four-dimensioned space of the mathematicians to the space from which the spirit-friends of the spiritualistic mediums entertain us with rappings, knockings, and bad English. Before taking this step we will first discuss the question of the real existence of a four-dimensional space, not deciding the question whether this space, if it really does exist, is inhabited by reasonable beings who consciously act upon the world in which we exist.

Among the reasons which are put forward to prove the existence of a four-dimensional space containing the world, the least reprehensible are those which are based on the existence of symmetry. We spoke above of two triangles in the same plane which have all

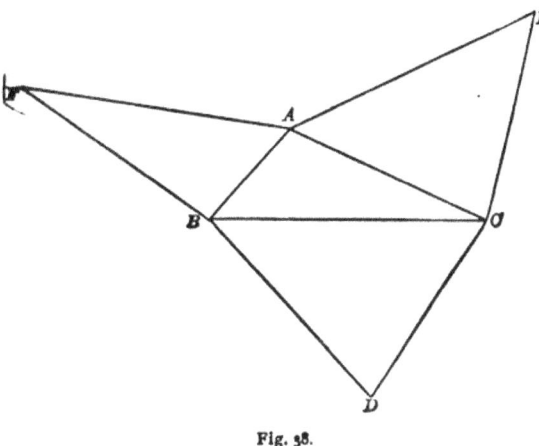

Fig. 38.

their sides and angles congruent, but which cannot be made to coincide by simple displacement within the plane ; but we saw that this coincidence could be effected by holding fast one side of one triangle and moving it out of its plane until it had been so far turned round that it fell back into its plane. Now something similar to this exists in space. Cut two figures, exactly like that of Fig. 38, out of a piece of paper, and turn the triangle $ABF$ about the side $AB$, $ACE$ about the side $AC$, $BCD$ about the side $BC$, and in one figure above and in the other below; then in both cases the points $D, E, F$ will meet at a point, because $AE$ is equal to $AF$, $BF$ is equal to $BD$, $CD$ is equal to $CE$. In this manner we obtain two pyramids which in all lengths and all angles are congruent, yet which cannot, no matter how we try, be made to coincide, that is, be so fitted the one into the other that they shall both stand as one pyramid. But the *reflected* image of the one could be brought into coincidence with the other. Two spatial structures whose sides and angles are thus equal to each other, and of which each may be viewed as the reflected image of the other, are called *symmetrical*. For instance, the right and the left hand are symmetrical; or, a right and a left glove. Now just as in two dimensions it is impossible by simple displacement to bring into congruence triangles which like those above mentioned can only be made to coincide by circumversion, so also in three dimensions it is impossible to bring into congruence two symmetrical pyramids. Careful mathematical reflection, however, declares that this could be effected, if it were possible, while holding one of the surfaces, to move the pyramid out of the space of experience, and to turn it round through a four-dimensioned space until it reached a point at which it would return again into our experiential space. This process would simply be the four-dimensional analogue of the three-dimensional circumversion in the above-mentioned case of the two triangles. Further, the interior surfaces in this process would be converted into exterior surfaces, and *vice versa*, exactly as in the circumversion of a triangle the anterior and posterior sides are interchanged. If the structure which is to be converted into its symmetrical counterpart is made of a flexible material, the interchange mentioned of the interior and exterior surfaces may be effected by simply turning the structure inside out; for example, a right glove may thus be converted into a left glove.

Now from this truth, that every structure can be converted, by means of a four-dimensional space inclusive of the world, into a structure symmetrical with it, it has been sought to establish the probability of the real existence of a four-dimensioned space. Yet it will be evident, from the discussion of the preceding section, that the only inference which we can here make is, that the idea of a four-dimensioned space is competent, from a mathematical point of view, to throw some light upon the phenomena of symmetry. To conclude from these facts that a space of this kind really exists, would be as daring as to conclude from the fact that the uniform

angular velocity of the apparent motions of the fixed stars is expli-
cable from the assumption of an axial motion of the firmament, that
the fixed stars are really rigidly placed in a celestial sphere rotating
about its axis.  It must not be forgotten that our comprehension of
the phenomena of the real world consists of two elements: first, of
that which the things really are;  and, second, of that by which we
rationally apprehend the things.   This latter element is partly dependent on the sum of the experiences which we have before acquired, and partly on the necessity, due to the imperfection of reason, of our classifying the multitudinous isolated phenomena of the
world into categories which we ourselves have formed, and which,
therefore, are not wholly derived from the phenomena themselves,
but to a great extent are dependent on us.

Besides geometrical reasons, Zöllner has also adduced cosmological reasons to prove the existence of a four-dimensional space.
To these reasons belong especially the questions, whether the number of the fixed stars is infinitely great, whether the world is finite
or infinite in extension, whether the world had a beginning or will
have an end, whether the world is not hastening towards a condition
of equilibrium or dead level by the universal distribution of its matter
and energy;  the problems, also, of gravitation and action at a distance;  and finally, the questions concerning the relations between
the phenomena in the world of sense-perception to the unknown
things-in-themselves.   All these questions which can be decided in
no definite sense, led Zöllner and his followers to the assumption
that a four-dimensioned space inclusive of the space of experience
must really exist.   But more careful reflection will show that this
assumption does not dispose of the difficulties but simply displaces
them into another realm.   Furthermore, even if four-dimensioned
space did unravel and make clear all the cosmological problems
which have bothered the human mind, still, its existence would not
be proved thereby; it would yet remain a mere hypothesis, designed
to render more intelligible to a being who can only make experiences
in a three-dimensional space, the phenomena therein which are full
of mystery to it.   A four-dimensioned space would in such case possess for the metaphysician a value similar to that which the ether

possesses for the physicist. Still more convincing than these cosmological reasons to the majority of men is the physio-psychological
reason drawn from the phenomena of vision which Zöllner adduces.
Into this main argument we will enter in more detail.

When we "see" an object, as we all know, the light which proceeds or is reflected therefrom produces an image on the retina of
our eye; this image is conducted to our consciousness by means of
the optic nerve, and our reason draws therefrom an inference respecting the object. When, now, we look at a square whose sides
are a decimetre in length, and whose centre is situated at the distance
of a metre from the pupil of our eye, an image is produced on the
retina. But exactly the same image will be produced there if we
look at a square whose sides are parallel to the sides of the first
square but two decimetres in length, and whose centre is situated
at a distance of two metres from the pupil of the eye. Proceeding
thus further, we readily discover that an eye can perceive in any
length or line only the ratio of its magnitude to the distance at which
it is situated from it, and that generally a three-dimensional world
must appear to the eye two-dimensional, because all points which
lie behind each other in the direction outwards from the eye produce on the retina only one image. This is due to the fact that the
retinal images are themselves two-dimensional; for which reason,
Zöllner says, the world must appear to a child as two-dimensional,
if it be supposed to live in a primitive condition of unconscious mental activity. To such a child two objects which are moving the one
behind the other, must appear as suffering displacement on a surface, which we conceive behind the objects, and on which the latter
are projected. In all these apparent displacements, coincidences
and changes of form also are effected. All these things must appear
puzzling to a human being in the first stages of its development,
and the mind thus finds itself, as Zöllner further argues, in the first
years of childhood forced to adopt a hypothesis concerning the constitution of space and to assume that the world is three-dimensional,
although the eye can really perceive it as only two-dimensional.
Zöllner then further says, that in the explanation of the effects of
the external world, man constantly finds this hypothesis of his child-

ish years confirmed, and that in this way it has become in his mind
so profound a conviction that it is no longer possible for him to
think it away.   Consonant with this argumentation, also, is Zöllner's remark, that the same phenomenon has presented itself in
astronomical methods of knowledge.  To explain the movements of
the planets, which  appear  to describe regular paths on the surface
of a celestial sphere, we were compelled in the solution of  the riddles which these motions presented, to assume in the structure of the
heavens a dimension of ''depth,'' and the  complicated motions in
the two-dimensioned firmament were converted into very simple
motions in three-dimensioned space.  Zöllner also contends that our
conception of the entire visible world as possessed of three dimensions is a product of our reason, which the mind was driven to form
by the contradictions which would be presented to it on the assumption of only two  dimensions by the perspective distortions, coincidences, and changes of magnitude of objects.  When a child moves
its hand before its eyes, turns it, brings it nearer, or pushes it farther
away, this child successively receives the most various impressions
on the surface of its retina of one and the same object of whose
identity and constancy its feelings offer it a perfect assurance.   If
the child regarded the changeable projection of the hand on the surface of the retina as the real object, and not the hand which lies beyond it, the child would constantly be met with contradictions in its
experience, and to avoid this it makes the hypothesis that the space
of experience is three-dimensional.   Zöllner's contention is, therefore, that man originally had only a two-dimensional intuition of
space, but was forced by experience to represent to himself the objects which on the retinal surface appeared two-dimensional, as
three-dimensional, and thus to transform his two-dimensional space-
intuition into a three-dimensional one.   Now, in exactly the same
way, according to Zöllner's notion, will man, by the advancement
and increasing exactness of his knowledge of the phenomena of the
outer world, also be compelled to conceive of the material world as
a ''shadow cast by a more real four-dimensional world,'' so that
these conceptions will be just as trivial for the people of the twen-
tieth century as since Copernicus's time the explanation of the mo-

tions of the heavenly bodies by means of a three-dimensional motion has been.

Zöllner's arguments from the phenomena of vision may be re-
futed as follows : In the first place it is incorrect to say that we see
the things of the external world by means of two-dimensional retinal
images.   The light which penetrates the eye causes an irritation
of the optic nerves, and any such effect which, though it be not
powerful, is, nevertheless, a mechanical one, can only take place on
things which are material.   But material things are always three-
dimensional.   The effect of light on the sensitive plates of photog-
raphy can with just as little justice be regarded as two-dimensional.
Our senses can have perception of nothing but three-dimensional
things, and this perception is effected by forces which in their turn
act on three-dimensional things, namely our sensory nerves.   It is
wrong to call an image two-dimensional, for it is only by abstraction
that we can conceive of a thickness so growing constantly smaller
and smaller as to admit of our regarding a three-dimensional picture
as two-dimensional, by giving it in mind a vanishingly small thickness.   It is also wrong to say, as Zöllner says, that when we see the
shadow of a hand which is cast upon a wall we see something two-
dimensional.   What we really perceive is that no light falls upon
our eye from the region included by the shadow, while from the
entire surrounding region light does fall on our eye.   But this light
is reflected from the material particles which form the surface of the
wall, that is, from three-dimensional particles of matter.   We must
always remember that our eye communicates to us only three-dimensional knowledge, and that for the comprehension of anything which
has two, one, or no dimensions, *a purely intellectual act of abstraction
must be added to the act of perception.*   When we imagine we have
made a lead-pencil mark on paper, we have, exactly viewed, simply
heaped alongside of each other little particles of graphite in such a
manner that there are by far fewer graphite particles in the lateral
and upward directions than there are in the longitudinal direction,
and thus our reason arrives by abstraction at the notion of a straight
line.   When we look at an object, say a cube of wood, we recognise
the object as three-dimensional, and it is only by abstraction that

we can conceive of its two-dimensional surfaces, of its twelve one-dimensional edges, and of its eight no-dimensional corners. For we reach the perception of its surface, for example, solely by reason of the fact that the material particles which form the cube prevent the transmission of light, and reflect it, whereby a part of the light reflected from every material particle strikes our eye. Now, by thinking exclusively of those material particles which are reflected, in contrariety to the empty space without and the hidden and therefore non-reflected particles within, we form the notion of a surface.

It is evident from this, that all that we perceive is three-dimensional, that we cannot reach anything two-dimensional without an intellectual abstraction, and that, therefore, we cannot conceive of anything two-dimensional exerting effects upon material things. But this fact is a refutation of the retinal argument of Zöllner. If vision consisted wholly and exclusively in the creation of a two-dimensional image, the things which take place in the world could never come into our consciousness. The child, therefore, does not originally apprehend the world, as Zöllner says, as two-dimensional; on the contrary, it apprehends it either not at all, or it apprehends it as three-dimensional. Of course the child must first "learn how" to see. It is found from the observation of children during the first months of their lives, and of the congenitally blind who have suddenly acquired the power of vision by some successful operation, that seeing does not consist alone in the irritations which arise in the optic nerves, but also in the correct interpretation of these irritations by reason. This correct interpretation, however, can be accomplished only by the accumulation of a considerable stock of experience. Especially must the recognition of the distance of the object seen be gradually learned. In this, two things are especially helpful; first, the fact that we have two eyes and, consequently, that we must feel two irritations of the optic nerves which are not wholly alike; and, secondly, the fact that we are enabled by our power of motion and our sense of touch to convince ourselves of the distance and form of the bodies seen. The question now arises, what sort of an intuition of space would a creature have that had only one eye, that could neither move itself nor its eye, and also possessed no peripheral nerves. According to Zöllner's view, this creature could, owing to its two-dimensional retinal images, have only a two-dimensional intuition of space. The author's opinion, however, is, that such a creature could not see at all, as it has no possibility of collecting experiences which are adapted in any way to interpreting the effects of things on its retina. The light which proceeded from the objects roundabout and fell on the retina could produce no other effect on the being than that of a wholly unintelligible irritation, or perhaps even pain.

The reflections presented sufficiently show that neither the phenomena of symmetry nor the retinal images of the objects of vision necessarily force upon us the assumption of a four-dimensioned space. If the material world should ever present problems which could not in the progress of knowledge be solved in a natural way, the assumption that a four-dimensional space containing the world exists would also be incompetent to resolve the difficulties presented; it would simply convert these difficulties into others, and not dispose of the problems but simply displace them to another world. Yet the question might be asked, is the existence of a four-dimensional space really *impossible?* To answer this question we must first clearly know what we mean by "exist." If existence means that the intellectual *idea* of a thing can be formed and that this idea shall not lead to contradictions with other well-established ideas and with experience, we have only to say that four-dimensioned space does exist, as the arguments adduced in sections III and IV have rendered plain. If, namely, the space of four dimensions did not exist as a clear idea in the minds of mathematicians, mathematicians could certainly not have been led by this idea to results which are recognised by the senses as true, and which really take place in our own representable space. But if existence means "material actuality," we must say that we neither now nor in the future can know anything about it. For we know material actuality only as three-dimensional, our senses can only make three-dimensional experiences, and the inferences of our reason, although they can well abstract from material things, can never ascend to the point of explaining a four-dimensional materiality. Just as little,

therefore, as we can locally fix the idea of a two-dimensional material world, as little can we ever verify the notion of a four-dimensional material existence.

### VI.

### EXAMINATION OF THE HYPOTHESIS CONCERNING THE EXISTENCE OF FOUR-DIMENSIONAL SPIRITS.

In connection with the belief that the visible world is contained
in a four-dimensioned space, Zöllner and his adherents further hold
that this higher space is inhabited by intelligent beings who can
act consciously and at will on the human beings who live in experiential space. To invest this opinion with greater strength, Zöllner
appealed to the fact that the greatest thinkers of antiquity and of
modern times were either wholly of this opinion or at least held
views from which his contentions might be immediately derived.
Plato's dialogue between Socrates and Glaukon in the seventh book
of the Republic, is evidence, says Zöllner, that this greatest philosopher of antiquity possessed some presentiment of this extension of
the notion of space. Yet any one who has connectedly studied and
understood Plato's system of philosophy must concede that the so-
called "ideas" of the Platonic system denote something wholly different from what Zöllner sees in them or pretends to see. Zöllner
says that these Platonic ideas are spatial objects of more than three
dimensions and represent "real existence" in the same sense that
the material world, as contrasted with the images on the retina, represents it. Zöllner similarly deals with the Kantian "thing-in-itself," which is also regarded as an object of higher dimensions.

To show Kant in the light of a predecessor, Zöllner quotes the
following passage from the former's "Träume eines Geisterseheis,
erläutert durch Träume der Metaphysik" (1766, *Collected Works*,
Vol. VII. page 32 et seq.): "I confess that I am very much in-
"clined to assert the existence of immaterial beings in the world,
"and to rank my own soul as one of such a class. It appears, there
"is a spiritual essence existent which is intimately bound up with
"matter but which does not act on those forces of the elements by
"which the latter are connected, but upon some internal principle

"of its own condition. It will, in the time to come—I know not
"when or where—be proved, that the human soul, even in this life,
"exists in a state of uninterrupted connection with all the imma-
"terial natures of the spiritual world; that it alternately acts on
"them and receives impressions from them, of which, as a human
"soul, it is not, in the normal state of things, conscious. It would
"be a great thing, if some such systematic constitution of the spirit-
"ual world, as we conceive it, could be deduced, not exclusively
"from our general notion of spiritual nature, which is altogether too
"hypothetical, but from some real and universally admitted ob-
"servations,—or, for that matter, if it could even be shown to be
"probable."

What Kant really asserts here is, first, the partly independent
and partly dependent existence of the soul, and of spiritual beings
generally, on matter, and, second, that spiritual beings have some
common connection with and mutually influence one another. This
contention, which is that of very many thinkers, does not, however, entail the consequence that the "transcendental subject of
Kant" must be four-dimensional, as Zöllner asserts it does. Kant
never even hinted at the theory that the psychical features of the
world owe their connection with the material features to the fact
that they are four-dimensional and, therefore, include the three-
dimensional. Is it a necessary conclusion that if a thing exists and
is not three-dimensional, as is the case with the soul, it is therefore four-dimensional? Can it not in fact be so constituted that it
is wholly meaningless to speak of dimensions at all in connection
with it?

Yet still more strangely than the words of Plato and Kant do
certain utterances of the mathematicians Gauss and Riemann speak
in favor of Zöllner's hypothesis. S. v. Waltershausen relates of
Gauss in his *Gruss zum Gedächtniss*, (Leipsic, 1856), that Gauss
had often remarked that the three dimensions of space were only
a specific peculiarity of the human mind. We can think ourselves,
he said, into beings who are conscious of only two dimensions;
similarly, perhaps, beings who are above and outside our world may
look down upon us; and there were, he continued, in a jesting tone,

a number of problems which he had here indefinitely laid aside, but hoped to treat in a superior state by superior geometrical methods. Leaving aside this jest, which quite naturally suggested itself, the remarks of Gauss are quite correct. We possess the power to abstract and can think, therefore, what kind of geometry a being that is only acquainted with a two-dimensional world would have; for instance, we can imagine that such a being could not conceive of the possibility of making two triangles coincide which were congruent in the sense above explained, and so on. So, also, we can understand that a being who has control of four dimensions can only conceive of a geometry of four-dimensional space, yet may have the capacity of thinking itself into spaces of other dimensions. But it does not follow from this that a four-dimensional space exists, let alone that it is inhabited by reasonable beings.

Riemann, on the other hand, speaks directly of a world of spirits. In his *Neue mathematische Principien der Naturphilosophie* he puts forth the hypothesis that the space of the world is filled with a material that is constantly pouring into the ponderable atoms, there to disappear from the phenomenal world. In every ponderable atom, he says, at every moment of time, there enters and appears a determinate amount of matter, proportional to the force of gravitation. The ponderable bodies, according to this theory, are the place at which the spiritual world enters and acts on the material world. Riemann's world of spirits, the sole office of which is to explain the phenomenon of gravitation as a force governing matter, is, however, essentially different from the spiritual world of Zöllner, the function of which is to explain supposed supersensuous phenomena which stand in the most glaring contradiction with the established known laws of the material world.

Besides this appeal to the testimony of eminent men like Plato, Kant, Gauss, and Riemann, the scientific prophet of modern spiritualism also bases his theory on the belief, which has obtained at all times and appeared in various forms among all peoples, that there exist in the world forces which at times are competent to evoke phenomena that are exempt from the ordinary laws of nature. We have but to think of the phenomena of table-turning which once excited the Chinese as much as it has aroused, during the last few decades, the European and American worlds; or of the divining-rod, by whose help our forefathers sought for water, in fact, as we do now in parts of Europe and America.

Cranz, in his essay on the subject, divides spiritualistic phenomena into physical and intellectual. Of the first class he enumerates the following: the moving of chairs and tables; the animation of walking-sticks, slippers, and broomsticks; the miraculous throwing of objects; spirit-rappings (Luther heard a sound in the Wartburg, "as if three score casks were hurled down the stairs"); the ecstatic suspension of persons above the floor; the diminution of the forces of gravity; the ordeals of witches; the fetching of wished-for objects; the declination of the magnetic needle by persons at a distance; the untying of knots in a closed string; insensibility to injury and exemption therefrom when tortured, as in handling red-hot coals, carrying hot irons, etc.; the music of invisible spirits; the materialisations of spirits or of individual parts of spirits (the footprints in the experiments of Slade, photographed by Zöllner); the double appearance of the same person; the penetration of matter (of closed doors, windows, and so forth). As numerous also is the selection presented by Cranz of intellectual phenomena, namely: spirit-writing (Have's instrument for the facilitation of intercourse with spirits), the clairvoyance and divination of somnambulists, of visionary, ecstatic, and hypnotised persons, prompted or controlled by narcotic medicines, by sleeping in temples, by music and dancing, by ascetic modes of life and residence in barren localities, by the exudations of the soil and of water, by the contemplation of jewels, mirrors, and crystal-pure water, and by anointing the finger-nails with consecrated oil. Also the following additional intellectual phenomena are cited: increased eloquence or suddenly acquired power of speaking in foreign languages; spirit-effects at a distance; inability to move, transferences of the will, and so forth.

All these phenomena, presented with the aspect of truth, and associated more or less with trickery, self-deception, and humbug, are adduced by the spiritualists to substantiate the belief in a world of spirits which intentionally and consciously take part in the events of the material world, and to prove that these phenomena may be sufficiently and consistently explained by the effects of the activity of such a world. It is impossible for us to discuss and put to the test here the explanations of all these supersensuous phenomena. Anything and everything can be explained by spirits who act at will upon the world. There are only a few of these phenomena, namely, clairvoyance and Slade's experiments, whose explanations are so intimately connected with our main theme, the so-called fourth dimension, that they cannot be passed over.

First, with respect to clairvoyance, the American visionary Davis describes the experiences which he claims to have made in this condition, induced by "magnetic sleep," as follows :* "The sphere of my vision now began to expand. At first, I could only clearly discern the walls of the house. At the start they seemed to me dark and gloomy; but they soon became brighter and finally transparent. I could now see the objects, the utensils, and the persons in the adjoining house as easily as those in the room in which I sat. But my perceptions extended further still ; before my wandering glance, which seemed to control a great semi-circle, the broad surface of the earth, for hundreds of miles about me, grew as transparent as water, and I saw the brains, the entrails, and the entire anatomy of the beasts that wandered about in the forests of the Eastern Hemisphere, hundreds and thousands of miles from the room in which I sat." The belief in the possibility of such states of clairvoyance is by no means new. Alexander Dumas made use of it, for example, in his *Mémoires d'un médicin*, in which Count Balsamo, afterwards called Cagliostro, is said to possess the power to throw suitable persons into this wonderful condition and thus to find out what other persons at distant places are doing. Zöllner explains clairvoyance by means of the fourth dimension thus :

A man who is accustomed to viewing things on a plain is supposed to ascend to a considerable height in a balloon. He will there enjoy a much more extended prospect than if he had remained on the plain below, and will also be able to signal to greater dis-

---

* Quoted by Cranz.

tances.   The plain, that is, the two-dimensioned space, is accordingly viewed by him from points outside of the plain as "open" in all directions.   Exactly so, in Zöllner's theory, must three-dimensioned spaces appear, when viewed from points in four-dimensioned space, namely, as "open"; and the more so in proportion as the point in question is situated at a greater distance from the place of our body, or in proportion as the soul ascends to a greater height in this fourth dimension.   Zöllner thus explains clairvoyance as a condition in which the soul has displaced itself out of its three-dimensioned space and reached a point which with respect to this space is four-dimensionally situated and whence it is able to contemplate the three-dimensional world without the interference of intervening obstacles.

The following remark is to be made to this explanation.   The reason why we have a better and more extended view from a balloon than from places on the earth is simply this, that between the suspended balloon and the objects seen at a distance nothing intervenes but the air, and air allows the transmission of light, whereas, at the places below on the earth there are all kinds of material things about the observer which prevent the transmission of light and either render difficult or absolutely impossible the sight of things which lie far away.   In the same way, also, from a point in four-dimensioned space, a three-dimensional object will be visible only provided there are no obstacles intervening.   If, therefore, this awareness of a distant object is a real, actual vision by means of a luminous ray which strikes the eye, there is contained in the explanation of Zöllner the tacit assumption that the medium with which the four-dimensional world is filled is also pervious to light exactly as the atmosphere is.

The theory that there are four-dimensional spirits who produce the phenomena cited by the spiritualists received special support from the experiments which the prestidigitateur Slade, who claimed he was a spiritualistic medium, performed before Zöllner.   Of these experiments we will speak of the two most important, the experiment with the glass sphere and the experiment with the knots.   To explain the connection of the glass sphere experiment with the fourth

dimension, we must first conceive of two-dimensional reasoning beings, or, let us say, two-dimensional worms, living and moving in a
plane. For a creature of this kind it will be self-evident that there
are no other paths between two points of its plane than such as lie
within the plane. It must, accordingly, be beyond the range of
conception of this worm, how any two-dimensional object which lies
within a circle in its space can be brought to any other position in
its space outside the circle without the object passing through the
barriers formed by the circle's circumference. But if this worm
could procure the services of a three-dimensional being, the trans-
portation of the object from a position within the circle to any position outside it could be effected by the three-dimensional being simply taking the object *out of* the plane and placing it at the desired
point. This object, therefore, would, in an inexplicable manner,
suddenly disappear before the eyes of the worms who were assem-
bled as spectators, and after the lapse of an interval of time would
again appear outside the circle without having passed at any point
through the circle's circumference. If now we add another dimension, we shall derive from this trick, which is wholly removed from
the sense-perception of the flattened worms, the following experiment, which is wholly beyond the perception of us human beings.
Inside a glass sphere, which is closed all around, a grain of corn is
placed; the problem is to transport the corn to some place outside
the sphere without passing through the glass surface. Now we
should be able to perform this trick if some four-dimensional being
would render us the same aid that we before rendered the two-
dimensional worm. For the four-dimensional being could take the
grain of corn into his four-dimensional space and then replace it in
our space in the desired spot outside of the glass sphere. Slade
performed this trick before Zöllner. Its mere performance sufficed
to convince this latter investigator that Slade had here made use of
a four-dimensional agent, who, in respect of power of motion, controlled his four-dimensional space as we do our three-dimensional
space. It never occurred to Zöllner that this experiment was the
cleverly executed trick of a prestidigitateur, or, as it would at once
occur to us, that the whole thing was a sensory illusion. The fact

that we cannot explain a trick easily and naturally does not irrevocably prove that it is accomplished by other means than those which the world of matter presents.

Still better known than this last performance is Slade's experiments with knots. To explain this in connection with the fourth dimension, we must resort again to the plane and the flat worm inhabiting it. To two parallel lines in a plane let the two ends of a third line, which has a double point, that is, intersects itself once, be attached. Our flat worm would not be able to untie the loop formed by the doubled third line, which we will call a string, because it cannot execute motions in three dimensions. If, therefore, a two-dimensional prestidigitateur should appear and accomplish the trick of untying this loop without removing the two ends of the string from the parallel lines, he will have accomplished for our flat worm a supersensuous experiment. A human being engaged in the service of the prestidigitateur could execute for him the experiment by simply lifting the string a little out of the plane, pulling it taut, and placing it back again. This suggests the following analogous experiment for three-dimensional beings. The two ends of a string in which a common knot has been made are sealed to the opposite walls of a room. The problem is to untie this knot without breaking the seals at the two ends of the string. Everybody knows that this problem is not soluble, but it may be calculated mathematically that the knot in the string can be untied as easily by motions in a fourth dimension of space as in the experiment above described the knot in the two-dimensional string was untied by a three-dimensional motion. Now as Slade untied the knot before Zöllner's eyes without apparently making any use of the ends fastened in the walls, Zöllner was still more firmly confirmed in the view that Slade had power over spirits who performed the experiments for him.

Still more far-reaching is the theory of Carl du Prel concerning the relations of the material and the four-dimensional world. (Compare his numerous essays in the spiritualistic magazine *Sphinx*.) Just as the shadows of three-dimensional objects cast on a wall are controlled in their movements by the things whose projections they are, in the same way it is claimed does there exist back of everything of this sense-perceptible world a real transcendental and four-dimensional "thing-in-itself" whose projection in the space of experience is what we falsely regard as the independent thing. Thus every man besides existing in his terrestrial self also exists in a spiritual or astral self which constantly accompanies him in his walks through life and whose existence is especially proclaimed in states of profound sleep, of somnambulism, and in the conditions of mediums. In this way Du Prel explains the wonderful feats of somnambulists, and the aerial journeys of sorcerers and witches. Whereas, ordinarily the separation of the material body from the astral body is only effected at the moment of death; in the case of somnambulists this separation may take place at any time, or, as Du Prel says, "the threshold of feeling may be permanently displaced."

In view of the natural relations of such theories to the dogmas of Christianity it is explainable that theologians also have raised their voices for or against spiritualism. While the *Protestant Church Times* beheld in the "repulsive thaumaturgic performances which these coryphæi of modern science offer, a lack of comprehension of real philosophy," the magazine *The Proof of Faith* expresses its delight at the discovery of spiritualism in the following manner: "Every Christian will surely rejoice at the deep and perhaps mortal wound which these new discoveries have in all probability administered to modern materialism."

We shall pass by the childish opinion that the Bible also speaks of four dimensions, as both in Job (xi. 8–9) and in the Epistle to the Ephesians (iii. 18) only breadth, length, depth, and height, that is, four directions of extension, are mentioned. Yet we will still add, as Cranz has done, the reflections which Zöllner, as the most prominent representative of modern spiritualism, has put forward respecting its relations to the doctrines of Christianity (*Wissensch. Abhandl.*, Vol. III). By the foundation of transcendental physics on the basis of spiritualistic phenomena, the "new light" has arisen which is spoken of in the New Testament. The rending of the veil of the Temple on the crucifixion of Christ, the resurrection, the ascension, the transfiguration, the speaking with many tongues on the giving out of the Holy Ghost, all these are in Zöllner's view spiritualistic phenomena. Similarly, he sees a reference to the four-dimensional world of spirits in all those sayings of Christ in which the latter speaks to his Apostles of the impossibility of their having any image or notion of the place to which when he disappeared he would go and whence he would return. (Gospel of St. John, xii. 33, 36; xiv. 2, 3, 28; xvi. 5, 13.)

Ulrici, however, goes farthest in the mingling of spiritualistic and Christian beliefs; for he sees in the doctrine of spiritualism a means of strengthening belief in a supreme moral world-order and in the immortality of the soul. In answer to Ulrici's tract "Spiritualism So-called, a Question of Science" (1889) Wundt wrote an annihilating reply bearing the title "Spiritualism, a Question of Science So-called." Wundt criticises the future condition of our soul according to spiritualistic hypotheses in the following sarcastic yet pertinent words, which Cranz also quotes: "(1) Physically, the "souls of the dead come into the thraldom of certain living beings "who are called mediums. These mediums are, for the present at "least, a not widely diffused class, and they appear to be almost "exclusively Americans. At the command of these mediums, de-"parted souls perform mechanical feats which possess throughout "the character of absolute aimlessness. They rap, they lift tables "and chairs, they move beds, they play on the harmonica, and do "other similar things. (2) Intellectually, the souls of the dead "enter a condition which, if we are to judge from the productions "which they deposit on the slates of the mediums, must be termed "a very lamentable one. These slate-writings belong throughout "in the category of imbecility; they are totally bereft of any con-"tents. (3) The most favored, apparently, is the moral condition of "the soul. According to the testimony which we have, its charac-"ter cannot be said to be anything else than that of harmlessness. "From brutal performances, such, for instance, as the destruction "of bed-canopies, the spirits most politely refrain." Wundt then laments the demoralising effect which spiritualism exercises on people who have hitherto devoted their powers to some serious pursuit or even to the service of science. In fact it is a presumptuous and flagrant procedure to forsake the path of exact research, which slow as it is, yet always leads to a sure extension of knowledge, in the hope that some aimless, foolhardy venture into the realm of uncertainty will carry us farther than the path of slow toil, and that we can ever thus easily lift the veil which hides from man the problems of the world that are yet unsolved.

*          *          *          .

Now that we have presented the opinions of others respecting the existence of a four-dimensional world of spirits, the author would like to develop one or two ideas of his own on the subject. In the preceding section it was stated that everything that we perceive by our senses is three-dimensional and that everything that possesses four or more dimensions can only be regarded as abstractions or fictions which our reason forms in its constant efforts after an extension and generalisation of knowledge. To speak of two-dimensional matter is as self-contradictory as the notion of four-dimensional matter. But a two or a four-dimensional world might exist in some other manner than a material manner, and for all we know in one which to us does not admit of representation. But in such a case, if it were without the power of affecting the material world, we should never be able to acquire any knowledge concerning its existence, and it would be totally indifferent to the people of the three-dimensional world, whether such a world existed or not. Just as an artist during his lifetime produces a number of different works of art, so also the Creator might have created a number of different-dimensioned worlds which in no wise interfere with one another. In such a case, any one world would not be able to know anything of any other, and we must consequently regard the question whether a four-dimensional world exists which is incapable of affecting ours, as insoluble. We have only to examine, therefore, the question whether the phenomenal world perhaps is a single individual in a great layer of worlds of which every successive one has one more dimension than the preceding and which are so connected with one another that each successive world contains and includes the preceding world, and, therefore, can produce effects in it. For our reason, which draws its inferences from the phenomena of this world, tells us, that if outside the three-dimensional world there exists a second four-dimensional world, containing ours, there is no reason why worlds of more or less dimensions should not, with the same right, also exist.   But if now, as Zöllner and his adherents maintain, four-dimensional spirits exist which can act by the mere efforts of their own wills on our world, there is surely no reason why we three-dimensional beings should not also be able to produce effects on some two-dimensional animated world.   Whether we have, generally, any such influence we do not know, but we certainly do know that we do not act purposely and consciously on a two-dimensional world.   If, therefore, Zöllner were right, the plan of creation would possess the wonderful feature that four-dimensional beings are capable of arbitrarily affecting the three-dimensional world, but that three-dimensional beings have no right in their turn consciously to affect a two-dimensional world.

The supporters of Zöllner's hypothesis will perhaps reply to the objection just made, that the plan of creation might, after all, possibly possess this wonderful peculiarity, that we human beings perhaps, in some higher condition of culture, will be able to act consciously on two-dimensional worlds, and that at any rate it is simply an inference by analogy to conclude from the non-existence of a relation between three and two dimensions that the same relation is also wanting in the case of four and three dimensions. As a matter of fact, the objection above made is not intended to refute Zöllner's hypothesis, but only to stamp it as very improbable.   But despite this improbability Zöllner would still be right if the phenomena of the material world actually made his hypothesis necessary. That, however, is by no means the case.   Although most of the phenomena to which the spiritualists appeal are probably founded on sense-illusions, humbug, and self-deception, it cannot be denied that there possibly do exist phenomena which cannot be brought into harmony with the natural laws now known.   There always have been mysterious phenomena, and there always will be.   Yet, as we have often seen that the progress of science has again and again revealed as natural what former generations held to be supernatural, it is certainly wholly wrong to bring in for the explanation of phenomena which now seem mysterious an hypothesis like that of Zöllner, by which

everything in the world can be explained.   If we adopt a point of
view which regards it as natural for spirits arbitrarily to interfere in
the workings of the world, all scientific investigation will cease, for
we could never more trust or rely upon any chemical or physical experiment, or any botanical or zoölogical culture.   If the spirits are
the authors of the phenomena that are mysterious to us, why should
they also not have control of the phenomena which are not mysterious?   The existence of mysterious phenomena justifies in no manner or form the assumption that spirits exist which produce them.
Would it not be much simpler, if we *must* have supernatural influences, to adopt the naïve religious point of view, according to
which everything that happens is traceable to the direct, actual, and
personal interference of a single being which we call God?   Things
which formerly stood beyond the sphere of our knowledge and were
regarded as marvellous events, as a storm, for example, now stand
in the most intimate connection with known natural laws.   Things
that formerly were mysterious are so no longer.   If one hundred and
fifty years ago some scientists were in the possession of our present
knowledge of inductional electricity and had connected Paris and
Berlin with a wire by whose aid one could clearly interpret in
Berlin what another person had at that very moment said in Paris,
people would have regarded this phenomenon as supernatural and
assumed that the originator of this long-distance speaking was in
league with spirits.

   We recognise, thus, that the things which are termed supernatural depend to a great extent on the stage of culture which humanity has reached.   Things which now appear to us mysterious,
may, in a very few decades, be recognised as quite natural.   This
knowledge, however, is not to be obtained by the lazy assumption
of bands of spirits as the authors of mysterious phenomena, but by
performing what in physics and chemistry is called experiment.
But the first and essential condition of all scientific experimenting
is that the experimenter shall be absolutely master of the conditions
under which the experiment is or is not to succeed.   Now, this cri-
terion of scientific experimenting is totally lacking in all spiritualistic
experiments.   We can never assign in their case the conditions un-

der which they will or will not succeed. When all the preparations in a spiritualistic *séance* have been properly made, but nothing takes place, the beautiful excuse is always forthcoming that the "spirits were not willing," that there were "too many incredulous persons present," and so forth. Fortunately, in physical experiments these pretexts are not necessary. By the path of experiment, and not by that of transcendental speculation, physics has thus made incredible progress and has piled new knowledge strata on strata upon the old. Accordingly, the prospect is left that the mysteries which the conditions and properties of the human soul still present can be solved more and more by the methods of scientific experiment. To this end, however, it is especially necessary that the physio-psychological experiments in question should only be performed by men who possess the critical eye of inquiry, who are free from the dangers of self-illusion, and who are competent to keep apart from their experiments all superstition and deception. The history of natural science clearly teaches that it is only by this road that man can arrive at certain and well-established knowledge. If, therefore, there really is behind such phenomena as mind-reading, telepathy, and similar psychical phenomena, something besides humbug and self-illusion, what we have to do is to study privately and carefully by serious experiments the success or non-success of such phenomena, and not allow ourselves to be influenced by the public and dramatic performances of psychical artists, like Cumberland and his ilk.

The high eminence on which the knowledge and civilisation of humanity now stands was not reached by the thoughtless employment of fanciful ideas, nor by recourse to four-dimensional worlds, but by hard, serious labor, and slow, unceasing research. Let all men of science, therefore, band themselves together and oppose a solid front to methods that explain everything that is now mysterious to us by the interference of independent spirits. For these methods, owing to the fact that they can explain everything, explain nothing, and thus oppose dangerous obstacles to the progress of real research, to which we owe the beautiful temple of modern knowledge.

# THE SQUARING OF THE CIRCLE.

## AN HISTORICAL SKETCH OF THE PROBLEM FROM THE RE-MOTEST TIMES.

I.

### UNIVERSAL INTEREST IN THE PROBLEM.

FOR two and a half thousand years, both competent and incompetent minds have striven in vain to solve the problem known as the squaring of the circle. Now that geometers have at last succeeded in giving a rigorous demonstration of the impossibility of solving the problem with straight edge and compasses, it seems fitting and opportune to cast a glance into the nature and history of this very ancient problem. And this will be found the more justifiable in view of the fact that the squaring of the circle, at least in name, is very widely known outside of the narrow circle of professional mathematicians.

The *Proceedings of the French Academy* for the year 1775 contain at page 61 a resolution of the Academy not to examine from that time on, any so-called solutions of the quadrature of the circle. The Academy was driven to this determination by the overwhelming multitude of professed solutions of the famous problem which were sent to it every month in the year,—solutions which of course were an invariable attestation of the ignorance and self-conceit of their authors, but which suffered collectively from the very important drawback in mathematics of being *wrong.* Since that time all professed solutions of the problem received by the Academy find a sure and permanent resting-place in the waste-basket, and remain unanswered for all time. The circle-squarer, however, sees in this high-handed manner of rejection only the envy of the great and powerful at his grand intellectual discovery. He is determined to secure recognition, and appeals therefore to the public. The newspapers must obtain for him the appreciation that scientific societies have denied. And every year the old mathematical sea-serpent more than once disports itself in the columns of our newspapers in the shape of an announcement that Mr. N. N., of P. P., has at last solved the problem of the quadrature of the circle.

But what manner of people are these circle-squarers, when examined by the light? Almost always they will be found to be imperfectly educated persons, whose mathematical knowledge does not exceed that of a modern high-school student. It is seldom that they know accurately what the requirements of the problem are and what its nature; they are totally ignorant of the two and a half thousand years' history of the problem; and they have no idea whatever of the important investigations which have been made with regard to it by great and real mathematicians in every century down to our own time.*

Yet great as is the quantum of ignorance that circle-squarers intermix with their intellectual products, the lavish supply of conceit and egotism with which they season their performances is still greater. I have not far to go to furnish a verification of this. A book printed in Hamburg in the year 1840 lies before me, in which the author thanks Almighty God at every second page that He has selected him and no one else to solve the "problem phenomenal" of mathematics, "so long sought for, so fervently desired, and attempted by millions." After this modest author has proclaimed himself the unmasker of Archimedes's deceit, he says: "And thus it hath pleased our mother Nature to withhold this precious mathematical jewel from the eye of human investigation, until she thought it fitting to reveal truth to simplicity."

This will suffice to show the great fatuity of the author. But it does not suffice to prove his ignorance. He has no conception

---

*For the full psychogeny and psychiatry of the circle-squarer see A. De Morgan, *A Budget of Paradoxes* (London, 1872). — *Tr.*

of mathematical demonstration ; he takes it for granted that things are so because they seem so to him.   Errors of logic, also, abound in his book.   But, minor fallacies apart, wherein does the real error of this "unmasker" of Archimedes consist?   It requires considerable labor to extricate the kernel of the demonstration from the turgid language and bombastic style in which the author has buried his conclusions.   But it is this.   The author inscribes a square in a circle, circumscribes another about it, then points out that the inside square is made up of four congruent triangles, whereas the circumscribed square is made up of eight such triangles ; from which fact, seeing that the circle is larger than the one square and smaller than the other, he draws the bold conclusion that the circle is equal in area to six such triangles.   It is hardly conceivable that a rational being could infer that something which is greater than 4 and less than 8 must necessarily be 6.   But with a man that attempts the squaring of the circle this kind of ratiocination *is* possible.

It is the same with all the other attempted solutions of the problem ; in all of them either logical fallacies or violations of elementary arithmetical or geometrical truths can be pointed out. Only they are not always of such a trivial nature as in the book just mentioned.

Let us now inquire into the origin of this propensity which leads people to occupy themselves with the quadrature of the circle.

Attention must first be called to the antiquity of the problem. A quadrature was attempted in Egypt 500 years before the exodus of the Israelites.   Among the Greeks the problem never ceased to play a part that greatly influenced the progress of mathematics. And in the middle ages also the squaring of the circle sporadically appears as the philosopher's stone of mathematics.   The problem has thus never ceased to be dealt with and considered.   But it is not by the antiquity of the problem that circle-squarers are enticed, but by the allurement which everything exerts that is calculated to raise the individual above the mass of ordinary humanity, and to bind about his temples the laurel crown of celebrity. Ambition spurred men on in ancient Greece and still spurs them on in modern times to crack this primeval mathematical nut.

Whether they are competent thereto is a secondary consideration. They look upon the squaring of the circle as the grand prize of a lottery that can just as well fall to their lot as that of any other man. They do not remember that—

"Toil before honor is placed by sagacious decrees of Immortals."

and that it requires years of consecutive study to gain possession of the mathematical weapons that are indispensably necessary to attack the problem, but which even in the hands of the most distinguished mathematical strategists did not suffice to take the stronghold.

But why is it, we must further ask, that it happens to be the squaring of the circle and not some other unsolved mathematical problem upon which the efforts of people are bestowed who have no knowledge of mathematics yet busy themselves with mathematical questions? The question is answered by the fact that the squaring of the circle is about the only mathematical problem that is known to the unprofessional world,—at least by name. Even among the Greeks the problem was very widely known outside of mathematical circles. In the eyes of the Grecian layman, as at present among many of his modern brethren, occupation with this problem was regarded as the most important and essential business of mathematicians. In fact, they had a special word to designate this species of activity, namely, τετραγωνίζειν, which means to busy one's self with the quadrature. In modern times, also, every educated person, though he be not a mathematician, knows the problem by name, and knows that it is insoluble, or at least, that despite the efforts of the most famous mathematicians it has not yet been solved. For this reason the phrase "to square the circle," is now generally used in the sense of attempting the impossible.

But in addition to the antiquity of the problem, and the fact also that it is known to the lay world, there is an important third factor that induces people to occupy themselves with it. This is the report that has been current for more than a century now, that the Academies, the Queen of England, or some other influential person has offered a large prize to be given to the one that first solves the problem. As a matter of fact, the hope of obtaining this large prize of money is with many circle-squarers the principal incitement to action. And the author of the book above referred to begs his readers to lend him their assistance in obtaining the prizes offered.

Although the opinion is widely current in the unprofessional world, that professional mathematicians are still busied with the solution of the problem, this is by no means the case. On the contrary, for some two hundred years, the endeavors of many great mathematicians have been directed solely towards demonstrating with exactness that the problem is insoluble. It is, as a rule,— and naturally,—more difficult to prove that a thing is impossible than to prove that it is possible. And thus it has happened, that up to within a few years ago, despite the employment of the most varied and the most comprehensive methods of modern mathematics, no one succeeded in supplying the wished-for demonstration of the problem's impossibility. At last, Professor Lindemann, of Königsberg, in June, 1882, succeeded in furnishing a demonstration,—and the first demonstration,—that it is impossible by employing only straight edge and compasses to construct a squaer that is mathematically exactly equal in area to a given circle. The demonstration, naturally, was not effected with the help of the old elementary methods; for if it were, it would have been accomplished centuries ago; but methods were requisite that were first furnished by the theory of definite integrals and departments of higher algebra developed in the last few decades; in other words, it required the direct and indirect preparatory labor of many centuries to make finally possible a demonstration of the insolubility of this historic problem.

Of course, this demonstration will have no more effect than the resolution of the Paris Academy of 1775, in causing the fecund race of circle-squarers to vanish from the face of the earth. In the future as in the past, there will be people who know nothing of this demonstration and will not care to know anything, and who believe that they cannot help succeeding in a matter in which others have failed, and that just they have been appointed by Providence to solve the famous puzzle. But unfortunately the ineradicable mania for solving the quadrature of the circle has also its serious side. Circle-squarers are not always so self-satisfied as the author of the book above mentioned. They often see, or at least divine, the insuperable difficulties that tower up before them, and the conflict between their aspirations and their performances, the consciousness that the problem they long to solve they are unable to solve, darkens their soul and, lost to the world, they become interesting subjects for the science of psychiatry.

## II.

### NATURE OF THE PROBLEM.

It is easy to determine the length of the radius of a circle, or the length of its diameter, which must be double that of the radius; and the question next arises, what is the number that tells how many times larger the circumference of the circle, that is the length of the circular line, is than its radius or its diameter. From the fact that all circles have the same shape it follows that this proportion will be the same for all circles both large and small. Now, since the time of Archimedes, all civilised nations that have cultivated mathematics have denoted the number that tells how many times larger the circumference of a circle is than the diameter by the symbol $\pi$, —the Greek initial letter of the word periphery.* To compute $\pi$, therefore, means to calculate how many times larger the circumference of a circle is than its diameter. This calculation is called "the numerical rectification of the circle."

Next to the calculation of the circumference, the calculation of the superficial contents of a circle by means of its radius or diameter is perhaps most important; that is, the computation of how great an area that part of a plane which lies within a circle measures. This calculation is called the "numerical quadrature." It depends, however, upon the problem of numerical rectification; that is, upon the calculation of the magnitude of $\pi$. For it is demonstrated in elementary geometry, that the area of a circle is

---

* The Greek symbol $\pi$ was first employed by W. Jones in 1706 and did not come into general use until about the middle of the eighteenth century through the works of Euler.—*Trans.*

equal to the area of a triangle produced by drawing in the circle a radius, erecting at the extremity of the same a tangent,—that is, in this case, a perpendicular,—cutting off upon the latter the length of the circumference, measuring from the extremity, and joining the point thus obtained with the centre of the circle. It follows from this that the area of a circle is as many times larger than the square upon its radius as the number $\pi$ amounts to.

The numerical rectification and numerical quadrature of the circle based upon the computation of the number $\pi$ are to be clearly distinguished from problems that require a straight line equal in length to the circumference of a circle, or a square equal in area to a circle, to be *constructively* produced from its radius or its diameter; problems which might properly be called "constructive rectification" or "constructive quadrature." Approximately, of course, by employing an approximate value for $\pi$, these problems are easily solvable. But to solve a problem of construction in geometry, means to solve it with mathematical exactitude. If the value $\pi$ were exactly equal to the ratio of two whole numbers to each other, the constructive rectification would present no difficulties. For example, suppose the circumference of a circle were exactly $3\frac{1}{7}$ times greater than its diameter; then the diameter could be divided into seven equal parts, which could easily be done by the principles of planimetry with straight edge and compasses; then by prolonging to the amount of such a part a straight line exactly three times as long as the diameter, we should obtain a straight line exactly equal to the circumference of the circle. But as a matter of fact,—and this has actually been demonstrated,—there do not exist two whole numbers, be they ever so great, that exactly represent by their proportion to each other the number $\pi$. Consequently, a rectification of the kind just described does not attain the object desired.

It might be asked here, whether from the demonstrated fact that the number $\pi$ is not equal to the ratio of two whole numbers however great, it does not immediately follow that it is impossible to construct a straight line exactly equal in length to the circumference of a circle; thus demonstrating at once the impossibility of

solving the problem. This question is to be answered in the negative. For in geometry there can easily exist pairs of lines of which
the one can be readily constructed from the other, notwithstanding
the fact that no two whole numbers can be found to represent the
ratio of the two lines. The side and the diagonal of a square, for
instance, are so constituted. It is true the ratio of the latter two
magnitudes is nearly that of 5 to 7. But this proportion is not
exact, and there are in fact no two numbers that represent the ratio
exactly. Nevertheless, either of these two lines can be readily constructed from the other by employing only straight edge and compasses. This might be the case, too, with the rectification of the
circle ; and consequently from the impossibility of representing $\pi$
by the ratio between two whole numbers the impossibility of the
problem of rectification is not inferable.

The quadrature of the circle stands and falls with the problem
of rectification. This rests upon the truth above mentioned, that
a circle is equal in area to a right-angled triangle, in which one
side is equal to the radius of the circle and the other to the circumference. Supposing, accordingly, that the circumference of the circle
had been rectified, then we could construct this triangle. But every
triangle, as we know from plane geometry, can, with the help of
straight edge and compasses be converted into a square exactly
equal to it in area. So that, supposing the rectification of the circumference of a circle to have been successfully effected, a square
could be constructed that would be exactly equal in area to the
circle.

The dependence upon one another of the three problems of the
computation of the number $\pi$, the quadrature of the circle, and its
rectification, thus obliges us, in dealing with the history of the
quadrature, to regard investigations with respect to the value of $\pi$
and attempts to rectify the circle as of equal importance, and to
consider them accordingly.

We have used repeatedly in the course of this discussion the
expression "to construct with straight edge and compasses." It
will be necessary to explain what is meant by the specification of
these two instruments. When to a requirement in geometry to

construct a figure there are so large a number of conditions an-
nexed that the construction of only *one* figure or a limited number
of figures is possible in accordance with those conditions ; such a
full and stated requirement is called a problem of construction, or
briefly a problem.   When a problem of this kind is presented for
solution it is necessary to reduce it to simpler problems, already
recognised as solvable ; and since these latter depend in their turn
upon other, still simpler problems, we are finally brought back to
certain fundamental problems upon which the rest are based but
which are not themselves reducible to problems less simple.   These
fundamental problems are, so to speak, the lowermost stones of the
edifice of geometrical construction.   The question next arises as to
what problems may be properly regarded as fundamental ; and it
has been found, that the solution of a great part of the problems
that arise in elementary plane geometry rests upon the solution of
only five original problems.   They are :

1. The construction of a straight line that shall pass through
two given points.

2. The construction of a circle the centre of which is a given
point and the radius of which has a given length.

3. The determination of the point lying coincidently on two
given straight lines prolonged as far as necessary,—in case such a
point (point of intersection) exists.

4. The determination of the two points that lie coincidently
on a given straight line and a given circle,—in case such common
points (points of intersection) exist.

5. The determination of the two points that lie coincidently on
two given circles,—in case such common points (points of intersection) exist.

For the solution of the three last of these five problems the
eye alone is needed, while for the solution of the first two problems, besides pencil, ink, chalk, or the like, additional special instruments are required : for the solution of the first problem a
straight edge or ruler is most generally used, and for the solution
of the second a pair of compasses.   But it must be remembered
that it is no concern of geometry what mechanical instruments are

employed in the solution of the five problems mentioned. Geometry simply limits itself to the presupposition that these problems are solvable, and regards a complicated problem as solved if, upon a specification of the constructions of which the solution consists, no other requirements are demanded than the five above mentioned. Since, accordingly, geometry does not itself furnish the solution of these five problems, but rather exacts them, they are termed *postulates*.\* All problems of plane geometry are not reducible to these five problems alone. There are problems that can be solved only by assuming other problems as solvable which are not included in the five given ; for example, the construction of an ellipse, having given its centre and its major and minor axes. Many problems, however, possess the property of being solvable with the assistance of the above-formulated five postulates alone, and where this is the case they are said to be ''constructible with straight edge and compasses,'' or ''elementarily'' constructible.

After these general remarks upon the solvability of problems of geometrical construction, which an understanding of the history of the squaring of the circle makes indispensable, the significance of the question whether the quadrature of the circle is or is not solvable, that is elementarily solvable, will become intelligible. But the conception of elementary solvability only gradually took clear form, and we therefore find among the Greeks as well as among the Arabs endeavors, successful in some respects, that aimed at solving the quadrature of the circle with other expedients than the five postulates. We have also to take these endeavors into consideration, and especially so as they, no less than the unsuccessful efforts at elementary solution, have upon the whole advanced the science of geometry, and contributed much to the clarification of geometrical ideas.

----

\* Usually geometers mention only two postulates (Nos. 1 and 2). But since to geometry proper it is indifferent whether only the eye, or additional special mechanical instruments are necessary, the author has regarded it more correct in point of method to assume five postulates.

III.

THE EGYPTIANS, BABYLONIANS, AND GREEKS.

In the oldest mathematical work that we possess we find a rule telling us how to construct a square which is equal in area to a given circle. This celebrated book, the Rhind Papyrus of the British Museum, translated and explained by Eisenlohr (Leipsic, 1877), was written, as stated in the work itself, in the thirty-third year of the reign of King Ra-a-us, by a scribe of that monarch, named Ahmes. The composition of the work falls accordingly in the period of the two Hyksos dynasties, that is, in the period between 2000 and 1700 B. C. But there is another important circumstance to be noted. Ahmes mentions in his introduction that he composed his work after the model of old treatises, written in the time of King Raenmat; whence it appears that the originals of the mathematical expositions of Ahmes are half a thousand years older still than the Rhind Papyrus.

The rule given in this papyrus for obtaining a square equal to a circle specifies that the diameter of the circle shall be shortened one-ninth of its length and upon the shortened line thus obtained a square erected. Of course, the area of a square of this construction is only approximately equal to the area of the circle. An idea may be obtained of the degree of exactness of this original, primitive quadrature by remarking, that if the diameter of the circle in question is one metre in length, the square that is supposed to be equal to the circle is a little less than half a square decimetre too large; an approximation not so accurate as that made by Archimedes, yet much more correct than many a one later employed. It is not known how Ahmes or his predecessors arrived at this approximate quadrature; but it is certain that it was handed down in Egypt from century to century, and in late Egyptian times it appears repeatedly.

In addition to the effort of the Egyptians, we also find in pre-Grecian antiquity an attempt at circle-computation among the Babylonians. This is not a quadrature, but is intended as a rectification of the circumference. The Babylonian mathematicians had discovered, that if the radius of a circle be successively inscribed as a chord within its circumference, after the sixth inscription we arrive at the point from which we set out, and they concluded from this that the circumference of a circle must be a little larger than a line which is six times as long as the radius, that is three times as long as the diameter. A trace of this Babylonian method of computation may even be found in the Bible; for in 1 Kings vii. 23, and 2 Chron. iv. 2, the great laver is described, which under the name of the "molten sea" constituted an ornament of the temple of Solomon; and it is said of this vessel that it measured ten cubits from brim to brim, and thirty cubits round about. The number 3 as the ratio of the circumference to the diameter is still more plainly given in the Talmud, where we read that "that which measures three lengths in circumference is one length across."

With regard to the earlier Greek mathematicians—as Thales and Pythagoras—we know that they acquired their elementary mathematical knowledge in Egypt. But nothing has been handed down to us which shows that they knew of the old Egyptian quadrature, or that they dealt with the problem at all. But tradition says, that, subsequently, the teacher of Euripides and Pericles, the great philosopher and mathematician Anaxagoras, whom Plato so highly praised, "drew the quadrature of the circle" in prison, in the year 434 B. C. This is the account of Plutarch in the seventeenth chapter of his work *De Exilio*. The method is not told us in which Anaxagoras is supposed to have solved the problem, and it is not said whether, knowingly or unknowingly, he gave an approximate solution after the manner of Ahmes. But at any rate, to Anaxagoras belongs the merit of having called attention to a problem that was to bear rich fruit by inciting Grecian scholars to busy themselves with geometry, and thus more and more to advance that science.

Again, it is reported that the mathematician Hippias of Elis invented a curved line that could be made to serve a double purpose: first, to trisect an angle, and second to square the circle. This curved line is the τετραγωνίζουσα so often mentioned by the later Greek mathematicians, and by the Romans called "quadrat-rix." Regarding the nature of this curve we have exact knowledge from Pappus. But it will be sufficient here to state that the quadratrix is not a circle nor a portion of a circle, so that its construction is not possible by means of the postulates enumerated in the preceding section. And therefore the solution of the quadrature of the circle founded on the construction of the quadratrix is not an elementary solution in the sense discussed in the last section. We can, it is true, conceive a mechanism that will draw this curve as well as compasses draw a circle; and with the assistance of a mechanism of this description the squaring of the circle is solvable with exactitude. But if it be allowed to employ in a solution an apparatus especially adapted thereto, every problem may be said to be solvable. Strictly taken, the invention of the curve of Hippias substitutes for one insuperable difficulty another equally insuperable. Some time afterwards, about the year 350 B. C., the mathematician Dinostratus showed that the quadratrix could also be used to solve the problem of rectification, and from that time on this problem plays almost the same rôle in Grecian mathematics as the related problem of quadrature.

As these problems gradually became known to the non-mathematicians of Greece, attempts at solution at once sprang up that are worthy of a place by the side of the solutions of modern amateur circle-squarers. The Sophists especially believed themselves competent by seductive dialectic to take the stronghold that had defied the intellectual onslaughts of the greatest mathematicians. With verbal nicety, amounting to puerility, it was said that the squaring of the circle depended upon the finding of a number which represented in itself both a square and a circle; a square by being a square number, a circle in that it ended with the same number as the root number from which, by multiplication with itself, it was produced. The number 36, accordingly, was, as they thought, the one that embodied the solution of the famous problem.

Contrasted with this twisting of words the speculations of Bryson and Antiphon, both contemporaries of Socrates, though inexact, appear in a high degree intelligent. Antiphon divided the circle into four equal arcs, and by joining the points of division obtained a square; he then divided each arc again into two equal parts and thus obtained an inscribed octagon; thence he constructed an inscribed 16-gon, and perceived that the figure so inscribed more and more approached the shape of a circle. In this way, he said, one should proceed, until there was inscribed in the circle a polygon whose sides by reason of their smallness should coincide with the circle. Now this polygon could, by methods already taught by the Pythagoreans, be converted into a square of equal area; and upon the basis of this fact Antiphon regarded the squaring of the circle as solved.

Nothing can be said against this method except that, however far the bisection of the arcs is carried, the result still remains an approximate one.

The attempt of Bryson of Heraclea was better still; for this scholar did not rest content with finding a square that was very little smaller than the circle, but obtained by means of circumscribed polygons another square that was very little larger than the circle. Only Bryson committed the error of believing that the area of the circle was the arithmetical mean between an inscribed and a circumscribed polygon of an equal number of sides. Notwithstanding this error, however, to Bryson belongs the merit—first, of having introduced into mathematics by his emphasis of the necessity of a square which was too large and one which was too small, the conception of upper and lower "limits" in approximations; and secondly, by his comparison of the regular inscribed and circumscribed polygons with a circle, of having indicated to Archimedes the way by which an approximate value of $\pi$ was to be reached.

Not long after Antiphon and Bryson, Hippocrates of Chios treated the problem, which had now become more and more famous, from a new point of view. Hippocrates was not satisfied with approximate equalities, and searched for curvilinearly bounded plane figures which should be mathematically equal to a rectilinearly bounded figure, and which therefore could be converted by straight edge and compasses into a square equal in area. First,

Hippocrates found that the crescent-shaped plane figure produced by drawing two perpendicular radii in a circle and describing upon the line joining their extremities a semicircle, is exactly equal in area to the triangle that is formed by this line of junction and the two radii; and upon the basis of this fact the endeavors of this un-tiring scholar were directed towards converting a circle into a crescent. Naturally he was unable to attain this object, but by his ef-forts he discovered many new geometrical truths; among others being the generalised form of the theorem mentioned, which bears to the present day the name of *lunulae Hippocratis*, the lunes of Hippocrates. Thus, in the case of Hippocrates, it appears in the plainest light, how precisely the insolvable problems of science are qualified to advance science; in that they incite investigators to devote themselves with persistence to its study and thus to fathom its utmost depths.

Following Hippocrates in the historical line of the great Gre-cian geometricians comes the systematist Euclid, whose rigid form-ulation of geometrical principles has remained the standard presen-tation down to the present century. The Elements of Euclid, however, contain nothing relating to the quadrature of the circle or to circle-computation. Comparisons of surfaces which relate to the circle are indeed found in the work, but nowhere a computa-tion of the circumference of a circle or of the area of a circle. This palpable gap in Euclid's system was filled by Archimedes, the greatest mathematician of antiquity.

Archimedes was born in Syracuse in the year 287 B. C., and devoted his life, which was spent in that city, to the mathematical and the physical sciences, which he enriched with invaluable con-tributions. He lived in Syracuse till the taking of the town by Marcellus, in the year 212 B. C., when he fell by the hand of a Ro-man soldier whom he had forbidden to destroy the figures he had drawn in the sand. To the greatest performances of Archimedes the successful computation of the number $\pi$ unquestionably be-longs. Like Bryson he started with regular inscribed and circum-scribed polygons. He showed how it was possible, beginning with the perimeter of an inscribed hexagon, which is equal to six radii, to obtain by calculation the perimeter of a regular dodecagon, and then the perimeter of a figure having double the number of sides of that, and so on. Treating, then, the circumscribed polygons in a similar manner, and proceeding with both series of polygons up to a regular 96-sided polygon, he discovered on the one hand that the ratio of the perimeter of the inscribed 96-sided polygon to the diameter was greater than $6336 : 2017\frac{1}{4}$, and on the other hand, that the corresponding ratio with respect to the circumscribed 96-sided polygon was smaller than $14688 : 4673\frac{1}{2}$. He inferred from this, that the number $\pi$, the ratio of the circumference to the diameter, was greater than the fraction $\frac{6336}{2017\frac{1}{4}}$ and smaller than $\frac{14688}{4673\frac{1}{2}}$. Reducing the two limits thus found for the value of $\pi$, Archimedes then showed that the first fraction was greater than $3\frac{10}{71}$, and that the second fraction was smaller than $3\frac{1}{7}$, whence it followed with certainty that the value sought for $\pi$ lay between $3\frac{1}{7}$ and $3\frac{10}{71}$. The larger of these two approximate values is the only one usually learned and employed. That which fills us with most astonishment in the case of Archimedes's computation of $\pi$, is, first, the great acumen and accuracy displayed by him in all the details of the computation, and secondly the unwearied perseverance which he exercised in calculating the limits of $\pi$ without the help of the Arabian system of numerals and the decimal notation. For it must be considered that at many stages of the computation what we call the extraction of roots was necessary, and that Archimedes could only by extremely tedious calculations obtain ratios that expressed approximately the roots of given numbers and fractions.*

With regard to the mathematicians of Greece that follow Archimedes, all refer to and employ the approximate value of $3\frac{1}{7}$ for $\pi$, without, however, contributing anything essentially new to the problems of quadrature and of cyclometry. Thus Hero of Alexandria, the father of surveying, who flourished about the year 100 B. C., employs for purposes of practical measurement some-

---

*For Archimedes's actual researches, see Rudio, *Archimedes, Huygens, Lambert, Legendre, vier Abhand. über die Kreismessung* (Leipsic, 1892), where translations of the works of these four authors on cyclometry will be found.—*Tr.*

times the value $3\frac{1}{7}$ for $\pi$ and sometimes even the rougher approximation $\pi = 3$. The astronomer Ptolemy, who lived in Alexandria about the year 150 A. D., and who was famous as being the author of the planetary system universally recognised as correct down to the time of Copernicus, was the only one who furnished a more exact value; this he designated, in the sexagesimal system of fractional notation which he employed, by 3, 8, 30,—that is 3 and $\frac{8}{60}$ and $\frac{30}{3600}$, or as we now say 3 degrees, 8 minutes (*partes minutae primae*), and 30 seconds (*partes minutae secundae*). As a matter of fact, the expression $3 + \frac{8}{60} + \frac{30}{3600} = 3\frac{17}{120}$ represents the number $\pi$ more exactly than $3\frac{1}{7}$; but on the other hand, is, by reason of the magnitude of the numbers 17 and 120 as compared with the numbers 1 and 7, more cumbersome.

IV.

THE ROMANS, HINDUS, CHINESE, ARABS, AND THE CHRISTIAN
NATIONS TO THE TIME OF NEWTON.

In the mathematical sciences, more than in any other, the Romans stood upon the shoulders of the Greeks. Indeed, with respect to cyclometry, they not only did not add anything new to the Grecian discoveries, but frequently even evinced that they either did not know of the beautiful result obtained by Archimedes, or at least could not appreciate it. For instance, Vitruvius, who lived during the time of Augustus, computed that a wheel 4 feet in diameter must measure $12\frac{1}{2}$ feet in circumference; in other words, he made $\pi$ equal to $3\frac{1}{8}$. And, similarly, a treatise on surveying, preserved to us in the Gudian manuscript of the library of Wolfenbüttel, contains the following instructions for squaring the circle: Divide the circumference of a circle into four parts and make one part the side of a square; this square will be equal in area to the circle. Apart from the fact that the rectification of the arc of a circle is requisite to the construction of a square of this kind, the Roman quadrature, viewed as a calculation, is more inexact even than any other known computation; for its result is that $\pi = 4$.

The mathematical performances of the Hindus were not only greater than those of the Romans, but in certain directions surpassed even those of the Greeks. In the most ancient source of the mathematics of India that we know of, the Culvasûtras, which date back to a little before our chronological era, we do not find, it is true, the squaring of the circle treated of, but the opposite problem is dealt with, which might fittingly be termed the circling of the square. The half of the side of a given square is prolonged in length one third of the excess of half the diagonal over half the side, and the line thus obtained is taken as the radius of the circle equal in area to the square. The simplest way to obtain an idea of the exactness of this construction is to compute how great $\pi$ would have to be if the construction were exactly correct. We find out in this way that the value of $\pi$ upon which the Indian circling of the square is based, is about from five to six hundredths smaller than the true value, whereas the approximate $\pi$ of Archimedes, $3\frac{1}{7}$, is only from one to two thousandths too large, and that the old Egyptian value exceeds the true value by from one to two hundredths.

Cyclometry very probably made great advances among the Hindus in the first four or five centuries of our era; for Aryabhatta, who lived about the year 500 after Christ, states, that the ratio of the circumference to the diameter is 62832 : 20000, an approximation that in exactness surpasses even that of Ptolemy. The Hindu result gives 3·1416 for $\pi$, while $\pi$ really lies between 3·141592 and 3·141593. How the Hindus obtained this excellent value is told by Ganeça, the commentator of Bhâskara, an author of the twelfth century. Ganeça says that the method of Archimedes was carried still farther by the Hindu mathematicians; that by continually doubling the number of sides they proceeded from the hexagon to a polygon of 384 sides, and that by the comparison of the circumferences of the inscribed and circumscribed 384-sided polygons they found that $\pi$ was equal to 3927:1250. It will be seen that the value given by Bhâskara is identical with the value of Aryabhatta. It is further worthy of remark that the earlier of these two Hindu mathematicians does not mention either the value $3\frac{1}{7}$ of Archimedes or the value $3\frac{17}{120}$ of Ptolemy, but that the later one knows of both values and especially recommends that of Archimedes as the most useful for practical applications. Strange to say, the good approximate value of Aryabhatta does not occur in Brahmagupta, the great Hindu mathematician who flourished in the beginning of the seventh century; but we find the curious information in this author that the area of a circle is exactly equal to the square root of 10 when the radius is unity. The value of $\pi$ as derivable from this formula,—a value from two to three hundredths too large,—has unquestionably arisen on Hindu soil. For it occurs in no Grecian mathematician; and Arabian authors, who were in a better position than we to know Greek and Hindu mathematical literature, declare that the approximation which makes $\pi$ equal to the square root of 10, is of Hindu origin. It is possible that the Hindu people, who were addicted more than any other to numeral mysticism, sought to find in this approximation some connection with the fact that man has ten fingers, and that accordingly ten is the basis of their numeral system.

Reviewing the achievements of the Hindus generally with respect to the problem of quadrature, we are brought to recognise that this people, whose talents lay more in the line of arithmetical computation than in the perception of spatial relations, accomplished as good as nothing on the purely geometrical side of the problem, but that the merit belongs to them of having carried the Archimedean method of computing $\pi$ several stages farther, and of having obtained in this way a much more exact value for it—a circumstance that is explainable when we consider that the Hindus are the inventors of our present system of numeral notation, possessing which they easily outdid Archimedes, who employed the awkward Greek system.

With regard to the Chinese, this people operated in ancient times with the Babylonian value for $\pi$, or 3; but they possessed knowledge of the approximate value of Archimedes at least since the end of the sixth century. Besides this, there appears in a number of Chinese mathematical treatises an approximate value peculiarly their own, in which $\pi = 3\frac{7}{50}$; a value, however, which notwithstanding it is written in larger figures, is no better than that of

Archimedes.    Attempts at the *constructive* quadrature of the circle are not found among the Chinese.

Greater were the merits of the Arabians in the advancement of mathematics; and especially in virtue of the fact that they preserved from oblivion the results of both Greek and Hindu research and handed them down to the Christian countries of the West. The Arabians expressly distinguished between the Archimedian approximate value and the two Hindu values, the square root of 10 and the ratio 62832:20000.    This distinction occurs also in Muhammed Ibn Musa Alchwarizmî, the same scholar who in the beginning of the ninth century brought the principles of our present system of numerical notation from India and introduced it into the Mohammedan world. The Arabians, however, did not 'study the numerical quadrature of the circle only, but also the constructive; for instance, an attempt of this kind was made by Ibn Alhaitam, who lived in Egypt about the year 1000 and whose treatise upon the squaring of the circle is preserved in a Vatican codex, which unfortunately has not yet been edited.

Christian civilisation, to which we are now about to pass, produced up to the second half of the fifteenth century extremely insignificant results in mathematics.    Even with regard to our present problem we have but a single important work to mention; the work, namely, of Frankos von Lüttich on the squaring of the circle, published in six books, but preserved only in fragments.    The author, who lived in the first half of the eleventh century, was probably a pupil of Pope Sylvester II., who was himself a not inconsiderable mathematician for his time and the author of the most celebrated geometrical treatise of the period.

Greater interest came to be bestowed upon mathematics, and especially on the problem of the quadrature of the circle, in the second half of the fifteenth century, when the sciences again began to revive.    This interest was principally aroused by Cardinal Nicolas de Cusa, a man highly esteemed for his astronomical and calendarial studies.    He claimed to have discovered the quadrature of the circle by employing only straight edge and compasses and thus attracted the attention of scholars to the historic problem.    People believed the famous Cardinal, and marvelled at his wisdom, until Regiomontanus, in letters written in 1464 and 1465 and published in 1533, rigorously demonstrated that the Cardinal's quadrature was incorrect. The construction of Cusa was as follows. The radius of a circle is prolonged a distance equal to the side of the inscribed square; the line so obtained is taken as the diameter of a second circle, and in the latter an equilateral triangle is described; then the perimeter of the latter is equal to the circumference of the original circle. If this construction, which its inventor regarded as exact, be considered as a construction of approximation, it will be found to be more inexact even than the construction resulting from the value $\pi = 3\frac{1}{7}$. For by Cusa's method $\pi$ would be from five to six thousandths smaller than it really is.

In the beginning of the sixteenth century a certain Bovillius appears, who also gave the construction of Cusa,—this time without notice. But about the middle of the sixteenth century a book was published which the scholars of the time at first received with interest. It bore the proud title *De Rebus Mathematicis Hactenus Desideratis*. Its author, Orontius Finaeus, represented that he had overcome all the difficulties that had ever stood in the way of geometrical investigators; and incidentally he also communicated to the world the "true quadrature" of the circle. His fame was short-lived. For soon afterwards, in a book entitled *De Erratis Orontii*, the Portuguese Petrus Nonius demonstrated that Orontius's quadrature, like most of his other professed discoveries, was incorrect.

In the succeeding period the number of circle-squarers so increased that we shall have to limit ourselves to those whom mathematicians recognise. And particularly is Simon Van Eyck to be mentioned, who towards the close of the sixteenth century published a quadrature which was so approximate that the value of $\pi$ derived from it was more exact even than that of Archimedes; and to disprove it the mathematician Peter Metius was obliged to seek a still more accurate value than $3\frac{1}{7}$. The erroneous quadrature of Van Eyck was thus the occasion of Metius's discovery that the ratio 355:113, or $3\frac{16}{113}$, varied from the true value of $\pi$ by less than one one-millionth, eclipsing accordingly all values hitherto obtained. Moreover, it is demonstrable by the theory of continued fractions, that, admitting figures to four places only, no two numbers more exactly represent the value of $\pi$ than 355 and 113.

In the same way the quadrature of the great philologist Joseph Scaliger led to refutations. Like most circle-squarers who believe in their discovery, Scaliger also was little versed in the elements of geometry. He solved the famous problem, however,—at least in his own opinion,—and published in 1592 a book upon it, which bore the pretentious title *Nova Cyclometria*, and in which the name of Archimedes was derided. The baselessness of his supposed discovery was demonstrated to him by the greatest mathematicians of his time; namely, Vieta, Adrianus Romanus, and Clavius.

Of the erring circle-squarers that flourished before the middle of the seventeenth century three others deserve particular mention, —Longomontanus of Copenhagen, who rendered such great services to astronomy, the Neapolitan John Porta, and Gregory of St. Vincent. Longomontanus made $\pi = 3\frac{14185}{1000000}$, and was so convinced of the correctness of his result as to thank God fervently, in the preface to his work *Inventio Quadraturae Circuli*, that He had granted him in his old age the strength to conquer the celebrated difficulty. John Porta followed the example of Hippocrates and endeavored to solve the problem by a comparison of lunes. Gregory of St. Vincent published a quadrature, the error of which was very hard to detect but was finally discovered by Descartes.

Of the famous mathematicians who dealt with our problem in the period between the close of the fifteenth century and the time of Newton, we first meet with Peter Metius, before mentioned, who succeeded in finding in the fraction 355:113 the best approximate value for $\pi$ involving small numbers only. The problem received a different advancement at the hands of the famous mathematician Vieta. Vieta was the first to whom the idea occurred of representing $\pi$ with mathematical exactness by an infinite series of definitely prescribed operations. By comparing inscribed and circumscribed polygons, Vieta found that we approach nearer and nearer to $\pi$ if we cause the operations of extracting the square root of $\frac{1}{2}$, and certain related additions and multiplications, to succeed each other in a certain manner, and that $\pi$ must come out exactly, if this series of operations could be continued indefinitely. Vieta thus found that to a diameter of 10000 million units a circumference belongs of from 31415 million 926535 units to 31415 million 926536 units of the same length.

But Vieta was outdone by the Netherlander Adrianus Romanus, who added five additional decimal places to the ten of Vieta. To accomplish this he computed with unspeakable labor the circumference of a regular circumscribed polygon of 1073741824 sides. This number is the thirtieth power of 2. Yet great as the labor of Adrianus Romanus was, that of Ludolf Van Ceulen was still greater; for the latter calculator succeeded in carrying the Archimedean process of approximation for the value of $\pi$ to 35 decimal places; that is, the deviation from the true value was smaller than one one-thousand quintillionth, a degree of exactness that we can have scarcely any conception of. Ludolf published the figures of the tremendous computation that led to his result. His calculation was carefully examined by the mathematician Griemberger and declared to be correct. Ludolf was justly proud of his work, and following the example of Archimedes, requested in his will that the result of his most important mathematical performance, the computation of $\pi$ to 35 decimal places, be engraved upon his tombstone ; a request which is said to have been carried out. In honor of Ludolf, $\pi$ is called to-day in Germany the Ludolfian number.

Although through the labor of Ludolf a degree of exactness for cyclometrical operations was now obtained that was more than sufficient for any practical purpose that could ever arise, neither the problem of constructive rectification nor that of constructive quadrature had been in any respect theoretically advanced thereby. The investigations conducted by the famous mathematicians and physicists Huygens and Snell about the middle of the seventeenth century, were more important from a mathematical point of view than the work of Ludolf. In his book *Cyclometricus* Snell took the position that the method of comparison of polygons, which originated with Archimedes and was employed by Ludolf, was not necessarily the best method of attaining the end sought ; and he succeeded by employing propositions which state that certain arcs of a circle are greater or smaller than certain straight lines connected with the circle, in obtaining methods that make it possible to reach results like the Ludolfian with much less labor of calculation. The beautiful theorems of Snell were proved a second time, and better proved, by the celebrated Dutch promoter of the science of optics, Huygens (*Opera Varia*, p. 365 et seq.; *Theoremata De Circuli et Hyperbolae Quadratura*, 1651), as well as perfected in many ways by him. Snell and Huygens were fully aware that they had advanced the problem of numerical quadrature only, and not that of the constructive quadrature. This plainly appeared in Huygens's case from the vehement dispute which he conducted with the English mathematician James Gregory. This controversy is significant for the history of our problem, from the fact that Gregory made the first attempt to prove that the squaring of the circle with straight edge and compasses was impossible. The result of the controversy, to which we owe many valuable tracts, was, that Huygens finally demonstrated in an incontrovertible manner the incorrectness of Gregory's proof of impossibility, adding that he also was of opinion that the solution of the problem with straight edge and compasses was impossible, but nevertheless was not himself able to demonstrate this fact. And Newton later expressed himself to the same effect. As a matter of fact a period of over 200 years elapsed before higher mathematics was far enough advanced to furnish a rigorous demonstration of impossibility.

### V.

### FROM NEWTON TO THE PRESENT.

Before we proceed to consider the promotive influence which the invention of the differential and the integral calculus exercised upon our problem, we shall enumerate a few at least of that never-ending succession of erring quadrators who delighted the world with the products of their ingenuity from the time of Newton to the present; and out of a pious and sincere regard for the contemporary world, we shall omit entirely to speak of the circle-squarers of our own time.

First to be mentioned is the celebrated English philosopher Hobbes. In his book *De Problematis Physicis*, in which he proposes to explain the phenomena of gravity and of ocean tides, he also takes up the quadrature of the circle and gives a very trivial construction, which in his opinion definitively solved the problem. It made $\pi = 3\frac{1}{5}$. In view of Hobbes's importance as a philosopher, two mathematicians, Huygens and Wallis, thought it proper to refute him at length. But Hobbes defended his position in a special treatise, where to sustain at least the appearance of being right, he disputed the fundamental principles of geometry and the theorem of Pythagoras.

In the last century France especially was rich in circle-squarers. We will mention: Oliver de Serres, who by means of a pair of scales determined that a circle weighed as much as the square upon the side of the equilateral triangle inscribed in it, that therefore they must have the same area, an experiment in which $\pi = 3$; Mathulon, who offered in legal form a reward of a thousand dollars to the person who would point out an error in his solution of the problem, and who was actually compelled by the courts to pay the money; Basselin, who believed that his quadrature must be right because it agreed with the approximate value of Archimedes, and who anathematised his ungrateful contemporaries, in the confidence that he would be recognised by posterity; Liger, who proved that a part is greater than the whole and to whom therefore the quadrature of the circle was child's play; Clerget, who based his solution upon the principle that a circle is a polygon of a definite number of sides, and who calculated, also, among other things, how large the point is at which two circles touch.

Germany and Poland also furnish their contingent to the army of circle-squarers. Lieutenant-Colonel Corsonich produced a quadrature in which $\pi$ equalled $3\frac{1}{5}$, and promised fifty ducats to the person who could prove that it was incorrect. Hesse of Berlin wrote an arithmetic in 1776, in which a true quadrature was also "made known," $\pi$ being exactly equal to $3\frac{11}{78}$. About the same time Professor Bischoff of Stettin defended a quadrature previously published by Captain Leistner, Preacher Merkel, and Schoolmaster

Böhm, which virtually made $\pi$ equal to the square of $\frac{4\frac{1}{2}}{}$, not even attaining the approximation of Archimedes.

From attempts of this character are to be clearly distinguished constructions of approximation in which the inventor is aware that he has not found a mathematically exact construction, but only an approximate one.   The value of such a construction will depend upon two things—first, upon the degree of exactness with which it is numerically expressed, and secondly on whether the construction can be easily made with straight edge and compasses.   Con-structions of this kind, simple in form and yet sufficiently exact for practical purposes, have been produced for centuries in great num-bers.   The great mathematician Euler, who died in 1783, did not think it out of place to attempt an approximate construction of this kind.   A very simple construction for the rectification of the circle and one which has passed into many geometrical text-books is that published by Kochansky in 1685 in the *Leipziger Berichte*.   It is as follows: "Erect upon the diameter of a circle at its extremities perpendiculars; with the centre as vertex and the diameter as side construct an angle of 30°; find the point of intersection of the line last drawn with the perpendicular, and join this point of inter-section with that point on the other perpendicular which is dis-tant three radii from the base of the perpendicular.   The line of junction so obtained is very approximately equal to one-half of the circumference of the given circle."   Calculation shows that the dif-ference between the true length of the circumference and the line thus constructed is less than $\frac{3}{100000}$ of the diameter.

Although such constructions of approximation are very inter-esting in themselves, they nevertheless play but a subordinate rôle in the history of the squaring of the circle; for on the one hand they can never furnish greater exactness for circle-computation than the thirty-five decimal places which Ludolf found, and on the other hand they are not adapted to advance in any way the ques-tion whether the exact quadrature of the circle with straight-edge and compasses is possible.

The numerical side of the problem, however, was considerably advanced by the new mathematical methods perfected by Newton and Leibnitz, and known as the differential and the integral calculus.

About the middle of the seventeenth century, before Newton and Leibnitz represented $\pi$ by series of powers, the English mathematicians Wallis and Lord Brouncker, Newton's predecessors in certain lines, succeeded in representing $\pi$ by an infinite series of figures combined according to the first four rules of arithmetic. A new method of computation was thus opened. Wallis found that the fourth part of $\pi$ is represented by the regularly formed product

$$\tfrac{2}{3} \times \tfrac{4}{3} \times \tfrac{4}{5} \times \tfrac{6}{5} \times \tfrac{6}{7} \times \tfrac{8}{7} \times \tfrac{8}{9} \times \text{ etc.}$$

more and more exactly the farther the multiplication is continued, and that the result always comes out too small if we stop at a proper fraction but too large if we stop at an improper fraction. Lord Brouncker, on the other hand, represents the value in question by a continued fraction in which the denominators are all 2 and the numerators are the squares of the odd numbers. Wallis, to whom Brouncker had communicated his elegant result without proof, demonstrated the same in his *Arithmetic of Infinites*.

The computation of $\pi$ could scarcely have been pushed to a greater degree of exactness by these results than that to which Ludolf and others had carried it by the older and more laborious methods. But the series of powers derived from the differential calculus of Newton and Leibnitz furnished a means of computing $\pi$ to hundreds of decimal places.

Gregory, Newton, and Leibnitz found that the fourth part of $\pi$ was equal exactly to

$$1 - \tfrac{1}{3} + \tfrac{1}{5} - \tfrac{1}{7} + \tfrac{1}{9} - \tfrac{1}{11} + \tfrac{1}{13} - \cdots$$

if we conceive this series, which is called the *Leibnitz series*, continued indefinitely. This series is wonderfully simple but is not adapted to the computation of $\pi$, for the reason that entirely too many members have to be taken into account to obtain $\pi$ accurately to a few decimal places only. The original formula, however, from which this series is derived, gives other formulæ which are excellently adapted to the actual computation. The original formula is the general series:

$$\alpha = a - \tfrac{1}{3}a^3 + \tfrac{1}{5}a^5 - \tfrac{1}{7}a^7 + \cdots,$$

where $\alpha$ is the length of the arc belonging to any central angle in a circle of radius 1, and $a$ the tangent to this angle. From this we derive the following :

$$\frac{\pi}{4} = (a+b+c+\ldots) - \tfrac{1}{3}(a^3+b^3+c^3+\ldots)$$
$$+ \tfrac{1}{5}(a^5+b^5+c^6+\ldots) - \ldots,$$

where $a$, $b$, $c \ldots$ are the tangents of angles whose sum is $45°$. Determining, therefore, the values of $a$, $b$, $c \ldots$, which are equal to small and convenient fractions and fulfil the conditions just mentioned, we obtain series of powers which are adapted to the computation of $\pi$.

The first to add by the aid of series of this description additional decimal places to the old 35 in the number $\pi$ was the English arithmetician Abraham Sharp, who, following Halley's instructions, in 1700 worked out $\pi$ to 72 decimal places. A little later Machin, professor of astronomy in London, computed $\pi$ to 100 decimal places, by putting, in the series given above, $a=b=c=d=\tfrac{1}{5}$ and $e = -\tfrac{1}{239}$; that is, by employing the following series :

$$\frac{\pi}{4} = 4 \cdot \left[ \frac{1}{5} - \frac{1}{3 \cdot 5^3} + \frac{1}{5 \cdot 5^5} - \frac{1}{7 \cdot 5^7} + \cdots \right]$$
$$- \left[ \frac{1}{239} - \frac{1}{3 \cdot 239^3} + \frac{1}{5 \cdot 239^5} - \cdots \right]$$

In the year 1819, Lagny of Paris outdid the computation of Machin, determining in two different ways the first 127 decimal places of $\pi$. Vega then obtained as many as 140 places, and the Hamburg arithmetician Zacharias Dase went as far as 200 places. The latter did not use Machin's series in his calculation, but the series produced by putting in the general series above given $a=\tfrac{1}{2}$, $b=\tfrac{1}{5}$, $c=\tfrac{1}{8}$. Finally, at a recent date, $\pi$ has been computed to 500 places.*

The computation to so many decimal places may serve as an illustration of the excellence of the modern methods as contrasted with those anciently employed, but it has otherwise neither a theo-

---

*In 1873 the approximation was carried by Shanks to 707 places of decimals. —*Trans.*

retical nor a practical value. That the computation of $\pi$ to say 15 decimal places more than sufficiently satisfies the subtlest requirements of practice may be gathered from a concrete example of the degree of exactness thus obtainable. Imagine a circle to be described with Berlin as centre, and the circumference to pass through Hamburg; then let the circumference of the circle be computed by multiplying its diameter by the value of $\pi$ to 15 decimal places, and then conceive it to be actually measured. The deviation from the true length in so large a circle as this even could not be as great as the 18 millionth part of a millimetre.

An idea can hardly be obtained of the degree of exactness produced by 100 decimal places. But the following example may possibly give us some conception of it. Conceive a sphere constructed with the earth as centre, and imagine its surface to pass through Sirius, which is 134½ millions of millions of kilometres distant from the earth. Then imagine this enormous sphere to be so packed with microbes that in every cubic millimetre millions of millions of these diminutive animalcula are present. Now conceive these microbes to be all unpacked and so distributed singly along a straight line, that every two microbes are as far distant from each other as Sirius from us, that is 134½ million million kilometres. Conceive the long line thus fixed by all the microbes, as the diameter of a circle, and imagine the circumference of it to be calculated by multiplying its diameter by $\pi$ to 100 decimal places. Then, in the case of a circle of this enormous magnitude even, the circumference so calculated would not vary from the real circumference by a millionth part of a millimetre.

This example will suffice to show that the calculation of $\pi$ to 100 or 500 decimal places is wholly useless.

Before we close this chapter upon the evaluation of $\pi$, we must mention the method, less fruitful than curious, which Professor Wolff of Zurich employed some decades ago to compute the value of $\pi$ to 3 places.* The floor of a room is divided up into equal squares, so as to resemble a huge chess-board, and a needle ex-

---

* See also A. De Morgan, *A Budget of Paradoxes*, pp. 169-171.—*Tr.*

actly equal in length to the side of each of these squares, is cast haphazard upon the floor. If we calculate, now, the probabilities of the needle so falling as to lie wholly within one of the squares, that is so that it does not cross any of the parallel lines forming the squares, the result of the calculation for this probability will be found to be exactly equal to $\pi - 3$. Consequently, a sufficient number of casts of the needle according to the law of large numbers must give the value of $\pi$ approximately. As a matter of fact, Professor Wolff, after 10000 trials, obtained the value of $\pi$ correctly to 3 decimal places.

Fruitful as the calculus of Newton and Leibnitz was for the evaluation of $\pi$, the problem of converting a circle into a square having exactly the same area was in no wise advanced thereby. Wallis, Newton, Leibnitz, and their immediate followers distinctly recognised this. The quadrature of the circle could not be solved; but it also could not be proved that the problem was insolvable with straight edge and compasses, although everybody was convinced of its insolvability. In mathematics, however, a conviction is only justified when supported by incontrovertible proof; and in the place of endeavors to solve the quadrature there accordingly now come endeavors to prove the impossibility of solving the celebrated problem.

The first step in this direction, small as it was, was made, by the French mathematician Lambert, who proved in the year 1761 that $\pi$ was neither a rational number nor even the square root of a rational number; that is, that neither $\pi$ nor the square of $\pi$ could be exactly represented by a fraction the denominator and numerator of which are whole numbers, however great the numbers be taken. Lambert's proof* showed, indeed, that the rectification and the quadrature of the circle could not be accomplished in one particular simple way, but it still did not exclude the possibility of the problem being solvable in some other more complicated way, and without requiring further aids than straight edge and compasses.

---

* Given in Legendre's *Geometry*, in the Appendix to De Morgan, *op. cit.*, p. 495, and in Rudio, *op. cit.*—*Tr*.

Proceeding slowly but surely it was next sought to discover the essential properties which distinguish problems solvable with straight edge and compasses from problems the construction of which is elementarily impossible, that is by employing the postulates only. Slight reflection showed, that a problem, to be elementarily solvable, must always be such that the unknown lines of its figure are connected with the known lines by an equation for the solution of which equations of the first and second degree only are requisite, and which can be so arranged that the measures of the known lines will appear as integers only. The conclusion to be drawn from this was that if the quadrature of the circle and consequently its rectification were solvable elementarily, the number $\pi$, which represents the ratio of the unknown circumference to the known diameter, must be the root of a certain equation, of a very high degree perhaps, but in which all the numbers are whole numbers; that is, there would have to exist an equation, made up entirely of whole numbers, which would be correct if its unknown quantity were made equal to $\pi$.

Since the beginning of this century, consequently, the efforts of a number of mathematicians have been bent upon proving that $\pi$ generally is not algebraical, that is, that it cannot be the root of an equation having whole numbers for coefficients. But mathematics had to make tremendous strides forward before the means were at hand to accomplish this demonstration. After the French Academician, Professor Hermite, had furnished important preparatory assistance in his treatise *Sur la Fonction Exponentielle*, published in the seventy-seventh volume of the *Comptes Rendus*, Professor Lindemann, at that time of Freiburg, now of Munich, finally succeeded, in June 1882, in rigorously demonstrating that the number $\pi$ is not algebraical,* and so supplied the first proof that the

---

*For the benefit of my mathematical readers I shall present here the most important steps of Lindemann's demonstration. M. Hermite in order to prove the transcendental character of

$$e = 1 + \frac{1}{1} + \frac{1}{1.2} + \frac{1}{1.2.3} + \frac{1}{1.2.3.4} + \ldots$$

developed relations between certain definite integrals (*Comptes Rendus* of the Paris Academy, Vol. 77, 1873). Proceeding from the relations thus established, problems of the rectification and squaring of the circle with the help only of algebraical instruments like straight edge and compasses are insolvable. Lindemann's proof appeared successively in the *Reports of the Berlin Academy* (June, 1882), in the *Comptes Rendus* of the French Academy (Vol. 115, pp. 72 to 74), and in the *Mathematische Annalen* (Vol. 20, pp. 213 to 225).

"It is impossible with straight edge and compasses to construct a square equal in area to a given circle." These are the words of the final determination of a controversy which is as old as the history of the human mind. But the race of circle-squarers, unmindful of the verdict of mathematics, the most infallible of arbiters, will never die out as long as ignorance and the thirst for glory remain united.

---

Professor Lindemann first demonstrates the following proposition : If the coefficients of an equation of the $n$th degree are all real or complex whole numbers and the $n$ roots of this equation $z_1$, $z_2$, ..., $z_n$ are different from zero and from each other it is impossible for

$$e^{z_1} + e^{z_2} + e^{z_3} \ldots + e^{z_n}$$

to be equal to $\frac{a}{b}$, where $a$ and $b$ are real or complex whole numbers. It is then shown that also between the functions

$$e^{rz_1} + e^{rz_2} + e^{rz_3} + \ldots e^{rz_n},$$

where $r$ denotes an integer, no linear equation can exist with rational coefficients different from zero. Finally the beautiful theorem results : If $z$ is the root of an irreducible algebraic equation the coefficients of which are real or complex whole numbers, then $e^z$ cannot be equal to a rational number. Now in reality $e^{\pi\sqrt{-1}}$ is equal to a rational number, namely, $-1$. Consequently, $\pi\sqrt{-1}$, and therefore $\pi$ itself, cannot be the root of an equation of the $n$th degree having whole numbers for coefficients, and therefore also not of such an equation having rational coefficients. The property last mentioned, however, $\pi$ would have if the squaring of the circle with straight edge and compasses were possible. [The questions involved in the discussions of the last three pages have been excellently treated by Klein in *Famous Problems of Elementary Geometry* recently translated by Beman and Smith (Ginn & Co., Boston). Lindemann's proof is here presented in a simplified form, and so brought within the comprehension of students conversant only with algebra.—*Tr.*]

# INDEX.

Ulrici, 107.
Ulysses, 84.
Unlimited, 85.
Unnamed number, 6.

Van Ceulen, Ludolf, 134.
Van Eyck, Simon, 132.
Van't Hoff, 88.
Vicenary system, 5,
Vieta, 133.
Vision, 93 et seq.
Vitruvius, 128.
Von Lüttich, Frankos, 131.

Wallis, 136, 138.
Waltershausen, 99.
Weierstrass, 31.
Wenzelides, 58.
Wislicenus, 88.
Wolfenbüttel Library, 128.
Wolff, Professor, 140.
Words, numeral, 4 et seq.
Worms, two-dimensional, 104.
Wundt, 107.

Zero, 13, 16.
Zero exponents, 21.
Zöllner, 65, 92, 93 et seq.

www.ingramcontent.com/pod-product-compliance
Lightning Source LLC
Chambersburg PA
CBHW080812180526
45168CB00006B/2414